Simplism

Simplism

The Art of Indirect Physics

Scott D. Bogart

Library of Congress Control Number: 2009913200
ISBN: Hardcover   978-1-4500-1121-1
                    Softcover   978-1-4500-1120-4
                    eBook   978-1-9845-7859-4

Print information available on the last page.

Rev. date: 05/06/2020

To order additional copies of this book, contact:
Xlibris
1-888-795-4274
www.Xlibris.com
Orders@Xlibris.com
585128

Indirect Physics

Indirect Physical Series

1. Collective Mass Theory 11

2. Creative Math and the Big Bang Equation 21

3. Terms and Definitions 49

4. ESP and AIDS Equivalents 81

5. Indirect Gartish Language 101

Preface

THE INDIRECT PHYSICAL series, as described under the title indirect physics, is not necessarily reluctant to physics in general or theoretical quantum physics concerning gravity. It is indeed the opposite. The series was to question everything with the sparked idea of pure mathematical thought based on questionable human error – whether biased or unbiased, coherent or incoherent – as long as the idea remains intact for further explanation(s) and to concern that through gravity, we can indirectly derive mechanics. And through nonmechanical subtlety, we can figure and equate the size and mass of the quark and all other elementary particles, their shape and everything else concerning the basic fundamentals of physics.

Also, the pure mathematical imagination expressed as certain brilliance comes to point out that we should indeed question all that is readily available to our human certainty. And because of our curious nature, we calculate to pursue a mathematical common sense of proof, which, by the results of intuition, we can advance into the final truth: a TOE.

It was the innate sense of error that gave me a deductive sense of learning. My errors, as I see them, as expressed in logical terms, helped me grasp the ideal physical properties of physics and the fun of science as a whole. So as to say, these are my indirect ideas of the entire spectrum as I saw them as a curious twenty-year-old: which, I hope, may contribute to the creative insight of readers and future generations.

"All of what is left is the geometrical structure and shape of the quark."

7

Note: In regard to the big bang, it appears that the creation of everything occurred at a rate exceeding that of the velocity of light. So, to question the calculated velocity when compared to the actual velocity, which may be different or correct, we must postulate the constant as a perfectly logical assumption of pure thought and imagination.

"To see the truth is to see the light."

The God Particle

200,000 mps

Indirect Physics

Collective Mass Theory

I N THE WORLD of physics, the idea is the secret to the stars. Within the idea, a secret is always contained. And within the secret lies an idea beyond all others. The ideas, upon which theories are built, become pursuant to our imagination. They capitalize on the wild and rare. These rarities captivate the unforgiving curiosity of possibilities never thought possible.

In the world of man power, there is an uncommon level of intelligence known as genius. Within the level of genius carries a fine line between itself and insanity. The fine line reassures the human mind of what is right and what is wrong. But if the human does not understand the fine line, then it is typically crossed numerous times within each day. And each day, the unrecognized retrospect of oneself involves ideas overlooked and marked insane. If the idea surfaces to society, then the influence of such an idea becomes a scary thought.

What is even scarier are the secrets to the ideas certain cultures chase. "Moving objects emit gravitational waves," stated Einstein. This prediction was one of many believed to be crazy. Something as rare as Einstein was typically misunderstood and stereotyped, but with patience, the believability became accepted. It takes the mind time to find the same thought pattern to trace its feasibility. This trace pattern, the common tread where most people do not roam, becomes frightening.

In ideas so rare to be cherished, quick but substantial guidance toward creativity becomes a must. The rendition of rarities becomes extremely diverse, and the influence overwhelms us beyond our wildest imagination. Through interpretations of each standard personality in which sixteen are defined, science can begin to understand the influence. And it only takes one idea to inspire the next. Thus begins the collective mass theory.

Physical Unity

Einstein stated that absolute time will never become possible to calculate. And in understood terms, the word *infinity* comes into its own. To declare finite or infinite is to pursue logic mathematically. And math, of all prevalent sciences, is an attempt to define everything in a theory known as the unified field.

The unified field combines similarities in force fields of electromagnetivity, the forces contained within atoms, and gravity. In the pursuit of the unified field, it is shown that common forces are being sewn together by showing similarities through simplicity. These similarities give hints and clues for the unity desired. Patterns define these ideas, and patterns are everything.

Our physical pattern lies within the mind due to the four hemispheres it contains. These hemispheres give the human the very functions we cannot live without. The persistent nature of curiosity captivates sequences within man to show representations, and the easiest representation in man is himself. Through these hemispheres, the easiest conscious physical pattern can be seen through duplication. When these are multiplied, they represent a physical pattern that can be taken one step further.

A duplicative measure using the *World Book Encyclopedia* indicates an intriguing number that shows psychological, as well as physical, unity. In our consciousness, it is possible to relate what is already present. For example, the highest mass in the periodic table represents the number 256, which can direct the attention of relative patterns. The sequence of physical unity calls for certain declarations of clarity through anything relevant. Finding these creative but claritive measures could resolve such high foundations of uniformity.

When the highest conscious mass is multiplied by itself to equal the highest mass understood in physics (equal to his number of personalities squared), it makes sense to pursue the overall unification through different fields of study as formats that relate in sequence that should not be changed until proper indication is stated otherwise. And if indication of creative oversight is not used, man should question his own faults.

Conscious Physical Pattern
(Psychological Unity)

$$4 \times 16 \times 256$$

1. 4 hemispheres of the mind
2. 16 number of personalities
3. 256 highest mass

Mistakes

The common decency of man is to understand himself before others so he does not collide in conflict. Understanding this gives us awareness. The terms of measurements in physics need to be questioned for what has been concluded.

Common sense is the decency that preludes the conclusiveness man craves for. Any miscalculation of misrepresentation that appears to make sense might be incorrect, even though people agree. There is a golden rule taught to young males. It takes a man to admit he is wrong, and it takes a man to retrace his steps to benefit the family. Our family is the math that gives us life. Life is as imperfect as the next, so does two pounds weight 1.987 pounds as expressed in units of uneven atomic masses on the periodic table?

The easiest and most feasible way to record mathematical measurements is to declare unity by evens when those evens are, in fact, even (not an odd number). Underlined math declares that two pounds weigh two pounds. The question is that the whole units of mass on the periodic table are questionable and have been miscalculated.

"Simplicity works if you work simplicity," states common sense. Simplified ideas, such as evens, make unification much easier. If the world were simplified, many more advances could be made and geared toward engineering, transportation, and health care. Common sense could be the final curse – common sense is everything.

Superconducting

Not long ago, United States was dedicating 11 billion dollars for a Superconducting Super Collider, or SSC, to resolve the hidden secrets within the atom that structuralize all elements found and how they function. The featured energy that could resolve large-scale questions about energy might have found the resolution to the hidden secrets inside man through the use of SSC and collisions of subatomic particles.

Once the collision of subatomic particles takes place, the high-energy mathematics becomes defined. The importance of these subatomic secrets could outline the forces that make unity possible. In doing so, it becomes important to realize how forces do and do not attract. The call for definitions of attractions through the SSC involves precautionary measures that commonly state when things collide, they divide unevenly. But with the development of probable error, physics has found that this is a difference science must face.

Superconductivity requires the collision of subatomic particles close to the speed of light. Preparations for calculations that require the velocity of light become questionable. In judging the measurement, it appears that the device designed for measuring the velocity had to use mirrors and other forms of diversionary tactics. Even the inclusion of gravity would seem to slow things down.

The overlooked becomes a miscalculation to the actual truth, and the word *miscalculation* may be shrewd upon and ignored because the error was never recognized. If math is derived from energy and energy – such as the velocity of light – is an even constant, it makes sense to believe light is an even number. Similarity of evens through precision of common sense also includes unity.

Simplicity

Even though science appears to be an abundant form of government funding, things change. This change calls for another extreme. This extreme is to brainstorm and figure out other diverse clues. Even listening to the most insane ideas gives science leads. When listening to these ideas, however, it becomes important that possibilities are taken into consideration, even though the person may appear highly insane. It is never known what each generation will bestow.

As generations come and go, the world always seems to find a way to make up for scientific ventures that become discontinued for added stimulation. Since having the SSC cancelled, it seems the time has come for the attempt of simplicity to resolve complexity (*Scientific American*, September, 40-47). Both extremes play a major role in determining what lies in between. Through extremes of simple measures, science can determine complex unification. The resolution would become a grand allure.

As things move toward resolution, examples exemplify our being. And while math keeps looking deep within, the more important examples become. One example, which exemplifies simplicity, would be to look toward someone who does the same. One of the most admired professional football players of all time, Joe Montana, as though it seems, could not pursue life or take new strides toward his own if his life were complicated. People often believe that a genius has a complex lifestyle. That is not necessarily true. As stated by his former president of the San Francisco 49ers, "Joe Montana's simplicity is his genius. He is able to operate on a simplistic level and come to decisions that other would think of as very complex." Simplicity is the true nature of greatness. Guys like Joe simply endure life with apparent ease (*GQ*). For so many, reality is the basic extreme – unbeatable in everyday life.

Resolution

High-energy studies are also here to question such ideas as Einstein's general theory of relativity. In the confusing world of today, it is hard to pinpoint legendary statutes that propel a person toward a fugitive nature.

In a diagram to express that moving objects emit gravitational waves, the expression is very simple. As an object moves across equal opposite mass – the prediction becomes reality. Through indirect physics, the attraction of positive and negative charges confirms the contingency.

> The atom is a moving object that attracts and emits
> opposing energy in all shapes and forms (gravity).

The resolution of an idea or theory leaves room for the next idea to surface and take hold as the previous one before. If the old is never replaced with the new, then generations will begin to conflict. Everyone wants to see someone take on the world through science with a new idea. It does not matter – it is exciting to see either way.

Each era needs something to look up to – not all ideas have complex resolutions. Some resolutions range accordingly on a Bell curve. The complex are few, as well as the simple, and the average mean is great. But if the range of the average is great – it operates on a scale not capable of being seen or construed until compiled. If shown, it is thus possible for "moving objects to emit gravitational waves."

> Even {{shaking}} a magnet will
> express Einstein's prediction.

Theory

So many times in this world, it is not necessarily the idea – instead it is what we make of it. Change takes a great amount of energy, and oftentimes, the energy it takes to imply change is so difficult that most people with good ideas back away, never recognizing their potential to benefit fields of study. But with persistence, from an educated evaluation, the sky is the limit.

Predictions give us imaginative ventures and liberal views. The pursuing variety of predictions add substituting specific terms to branch logic not previously recorded, and thus begins the influence of theory. The collective mass theory has a broad perspective. It includes what we have, and what we do not. The assumption that the universe is a multifaceted pursuit toward truth is a common cause. As the secrets of ourselves close in on the actual truth, somewhere lies the entirety.

Within the entirety, phenomena is found. The phenomena of education cause a spark to fly within ourselves and the universe. This spark is known as a bright idea. These ideas to justify become valued. In standard math, the theory attempts to define all. The exemplified following passage of the scientific method battles these ideas. The theory becomes looked upon, and it is almost certain for techniques to estimate or establish its feasibility. But when the idea is so far-fetched, it remains under the craziness scale and is never looked upon again. But even craziness can benefit things outside itself because even craziness has its fine points. Influence and greatness are judged by it.

Universe

The beginning of all creation is built upon a mass called the quark. The quark is found within every existence. This foundation of its size and weight has not yet been observed, but a search for the resolution is slowly being pursued.

The smallest particle, such as the quark, can be a very intriguing mystery. It appears that a more imaginative calculation can be achieved. One predictive calculation is a retrospective overview by canceling the human. And upon doing so, the evolution of creation and its genesis is canceled, and what is left is the smallest mass, the quark.

Another predictive theory foresees that man's numerical self could represent the final year before universal collapse. Anything is possible – divided by the energy fields that make it complete; so therefore, the possibility that both extremes could cancel to equal the existence who created could be universal truth.

Hidden truth is a concept that can recreate challenges for us to see. The outer edge of final human destiny is to fully contemplate the entirety of energy patterns known as the entire energy pattern theory. (The entire energy pattern theory could never become reality. Destiny is a question that is not consciously obtainable by human flesh – man does not live forever.)

Numerical Cancellation

The derivatives of man follow certain rules set forth by our Creator. As the rules unfold and become more complex, the more important they become to question man's purpose.

The numbers in the creative equation, five divided by five to the sixteenth power, represent man's numerical cancellation, the cancellation of the human. The expressed figure of $3.2768\text{-}10^{11}$ could, on a theoretical scale of standards and evens, be the size and mass of the quark. The cancellation may be a clue to where actual math came from because, as we all know, we count with both our hands and mind. This overview could be human destiny so truly searched for.

As we ponder, an inevitable question develops. This question of doubt and self-gratification arouses suspicion. What was meant to be was planned – so it might be feasible or possible for our Creator to use the same scale of measurement as us. The hint or clue might be for man to figure. The frenzy of speculation might be man's numeracies. And as the unified field theory states, similarities provide unity of extreme.

The personalities within man's numerical cancellation leave knowledge for conscious collectivity, which would be the highest power man has to calculate. This conscious collectivity of leaving man's mind outside of his digital power could fathom the mental energy to believe that absolute time is possible. Is it possible? Would this idea be a hint to the final piece of universal truth?

Periodic Time

As generations come and go, compiled research focused toward the periodic table becomes more and more valued. The expressed masses on the periodic table could also represent large quantum leaps of differing parallel universes. If numeracies are a hint to the truth, then time could equal weight.

Differing parallel universes are thought to be possible, so an idea or theory for imaginative ventures could be developed. As each universe reaches its maturity age, another develops.

The universal truth could be an elemental mass for each universe. As each universal mass is displayed, the overall age is equal to man's numerical cancellation. If one universe exists, it, too, contains the cancellation age. The complete set of universes equals the overall cancellation age, divided by the number of masses.

Example

$5/5^{16}$ + no. of masses (universes)

Periodic time ranges in degrees according to universal truth – and as the future rolls on, it is important for man to understand his own destiny because the future contains an abundance of questions that are always uncertain. Time travel and parallel universes might be the secrets that hold us from within.

Table Theory

256 universes, 103 masses

$$\longleftarrow \!\!\times\!\! \longrightarrow$$

Creation of the universe

Conclusion

As man continues to gain knowledge toward unity through math, he becomes determined to build a foundation. And if these equations are wrong, the complete entity evolves around the miscalculation. These common errors are lost because everything was centered around them.

Indirect Physics

Creative Math and the Big Bang Equation

A S THE FUTURE rolls on and becomes more highly complicated, new and intuitive theories are hard to come by. The duplicating measures, where ideals from one generation inspire the next, call for in-depth reasoning and fortitude. Once continuing duplication for furthering education develops, preparatory decisions suspend the ideas for everyone else to see. These basic complimentary cultures cherish the grace and beauty of such ingenuity that suggests common ideas can enhance the proposition. And the proposition of descriptive oversight can influence the outer world for continuing purposeful spontaneity. The idea of the quote "Simplicity works if you work simplicity" could bring us to that level.

Math is thought to be the answer to every problem described by man. The extremity of creating a new math to help enhance is not such a grave idea – but instead delightful. The familiarization of such new math calls for proof and numbers when the manner has never been approached. So, to approach a new math, one must be thorough and provide similarities of examples where any pertinent information may be found. In doing so, the mathematician becomes perceptive to his or her cause, while the math becomes judged and questioned for its feasibility.

Against any background, theoretical math can provide barriers that overwhelm man with knowledge. This suggested barrier gives us attitude and self-esteem to those who choose its influence. Through the viewpoint, the influenced can see the reasoning of its creation and consider acceptance.

Even agreeable measures as relating complexity with simplicity could ensue influence. As the examples of their very creation come into their own, they become duplicated, as done previously in every good idea. And as ideas develop, there is no limitation.

Psychological Physics

It is hard to pinpoint the exact nature of positive and negative attitudes. There are so many people with alternating backgrounds. We, as a human race, do possess a capability to break down these barriers through education.

Distinguishing the barriers begins with each personality and how it functions; better yet, the easiest way to inform others or yourself on what a positive or negative nature is would be to take a step back and look at things from a broad perspective. This perspective shows an intriguing combination of energies that can relate atomic charges of attraction and energy sharing.

1. When two negative people complain, they contain an attitude toward the subject. And once expressed, the energy appears to become a positive. Thus, when two people complain, the charge between them becomes positive. If written as an insight for atomic charge, it could be a positive.

2. When two highly optimistic people come together, they share a positive attitude that remains very constructive in its own right. And once they share the optimism, the energy of charge remains neutral from pessimism. The neutral base, from which people of positive attitudes refrain, gives society a strongpoint and keeps things intact. Thus, when two people share their attitude, they become neutral from the negative – two positives make a neutral.

3. When two very neutral people come together, they can be seen as not benefiting things outside themselves. So if two neutrals come together, the charge between them can be expressed as two neutrals equal a negative.

The possibilities of old and new combinations, such as derivatives in combining positive and negative for attraction, can plug in combinations useful in discovery. The inclusion of new discoveries can be quickly neutralized and rendered for its attraction. The attractions repel as well as convene. As a whole, these physical equivalents attract and neutralize. The attraction of positive and negative charges to equal a neutral is expressed as positive and negative make a neutral. And in physics, these could become a law if shown how to work properly.

Model Theory

The atomic charge expressed through psychological physics is an important role within the model theory 8, 4, 2. When the universe was created, a secret within every element, known as the quark, is found. The quark is the smallest particle. Its proper quark count has not yet been observed, but theories can give physics models to follow.

Everything reproduces, thus causing our human existence. Reproduction of mass can set an example of theory. The model theory that contains adequate quark counts under is the model theory 8, 4, 2. There are eight (8) per electron, four (4) per proton, and two (2) per neutron. Using psychological physics, science can derive its charges. The displayed cancellations used to deviate these charges, the atomic table.

The atomic table rules and format are as follows:

1. Use psychological physics / atomic charge.
2. Always cancel same charge.
3. Neutrals contained in subatomic units can influence overall charge.
4. Each subatomic unit contains a positive and a negative.

Atomic Table

Electron (N⁻)	Proton (N⁺)	Neutron (N)

The law of duplication states that "everything is equal to itself in a 2:1 ratio." The 2 represents the outside looking in (2:1) – sociology. The 1 represents the inside looking out – psychology. As the model theory 8, 4, 2 follows the same rules of duplication, everything can be declared a viewpoint from which both fields of study are made. Even the ratio can coherently break down arguments of viewpoints and install trust of comprehension.

Negative Space

In the model theory 8, 4, 2, the duplication of all opposing energies renders negative space. Simply put, everything with a boundary has negative space, just as though everything duplicates. (That is why we are here.) The missing number in this model issues a void of number 6. The negative space is the number that fills the void. And in filling a void, a boundary must exist. Within this model, these four are found: a void, negative space, the number 6, and, of course, a boundary. This completes all characteristics a model must have to be declared such.

Creative math can propose a digit to represent a boundary, such as using the number 1 for orbits, spins, distance, and boundaries. The best example is a prime example to show how a boundary of differing measures can be assessed. A ruler can formulate complete orbits or distance as a boundary of the ones using the number 11. If a boundary of an object ceases, the boundary of the two registers as one.

The boundary that contains negative space (6) is documented as number 16. Negative space is captured within the velocity that surrounds it. And atoms, otherwise known as elements, are the prime example and the foundation of all existence listed on the periodic table of elements. As these foundations of standards further develop our knowledge, it best be pointed out that in any math, negative space is relevant and should be based as the comprehensive distance between all objects.

Negative space is appreciated as outer space by most individuals. The enjoyment of its mystery engages curiosity toward the universe. If a name was to be given to the enigma, then people could relate on a simple level, such as language and communication. The enlightenment would be to see ourselves as being complete without lacking differences or loopholes of extremes without words. Communication of language breaks the void of negative space and calls attention where needed.

Negative space is a true element in itself. Even an equation can be devised to show its relevance. The center stage of nonexistence would be to declare negative space to have no mass; therefore, the atomic mass would be zero. The volume is bound by its captured velocity and rate of duplication. Inside every element, negative space is found. Even the universe is a captured velocity.

For every discovery, there is a discoverer who gets the right to name it how he sees fit. His right to label the discovery calls for attention. The discovered element of negative space, captured within all atoms and in which the universe presides around at a rate defined by a velocity of light, is equated as an infinite mass (∞v).

The infinite mass is as unlimited to man's comprehension as the elements themselves. Its nature of being a mass captured within a velocity hinders the foresight of name identification for communication barriers. It is named Bogartium, garth; element number 0; letter symbol B.

$$E = mc^2$$

Energy and its pattern has a peculiar nature. The creator was the inevitable Einstein. And even though overwhelming destruction occurred from his genius, humans have learned that simplicity is the prelude to the opposite spectrum.

The easiest way to calculate energy is to cancel the mass being understood as infinite mass or garth. With that in mind, the velocity of light squared can be calculated to equal infinite mass energy. Using the collective mass theory as a reference for creative math, squaring the velocity of light would then be 4 10. $E = mc^2$ becomes simplified. This philosophy can render certain energies that can be equated or duplicated.

As simplified creative math exemplifies the very character from which it was created, it leaves an unrelenting description for more statistics and their meaning. The descriptive nature of energy also relies on velocity and multicolored interaction of opposing forces.

The model theory 8, 4, 2 does have other possibilities. The negative space and its three charges found in its nucleus equal 36. To represent the highly figured number in itself through duplication and negative space equal to the number of personalities indicated by Myers-Briggs Type Indicator and an equation can be deviated by the standard of its boundary.

The precision cannot be potentially practiced if the deviants are not manageable. So, in governing the deviant of the smallest, it should be possible to trace energies of the highest. The atom's deviant is number 1.

"When cancelling man's numerical digit (5) to the power of its boundaries (11), the standard is expressed (0.0000001)."

It is important to ensue a calculating device in a time before a lot of math become complicated. The difference is that complexity distracts truth. And to calculate energy through creative math by the use of a computer until resolved would not be the true meaning. But the possibilities are endless once justified enough to follow.

Question: The insight of the big bang had a start of mass proportions. Is this model theory 8, 4, 2 the beginning of expanded matter?

Captured Mass

There are many different ways of calculating mass, but there always comes a time to show differing relationships in assessing its size and shape. The atom itself can be rightfully measured as a cube for its volume and size. It can be measured for its three dimensions: inside, outside, radius, or diameter. Understanding that man created the square, math can render the natural shape. A new technique to declare it a cube through measuring old techniques of a squared cube is used in capturing infinite mass. The volume is measured in a imaginary boxed cube that is big enough for each element. Once measured, the garth can be structured to emphasize a calculation.

The degree of measuring garth is attained by taking the actual mass and dividing it by 256, and vice versa for both inside and outside, until the negative space or volume is found as an error on the proper scientific calculator (Texas Instruments Scientific Calculator TI-30 III). This then registers the infinite mass as a cube for further uses of a paradigm. The name of the paradigm is equal to the Planck scale and its pyramid shape of five sides – trilateral triangular paradigm. The nonmechanical subtlety allows us to see all similarities of time and space, outlined by shape and size of strengths within mass to conclude unity in the unified field theory. These similarities allow us to see the same force fields of the weak force, electromagnetism, strong force, and gravity.

Trilateral

1. The element number is divided by the highest mass.
2. Divide the highest mass by the element number.
3. Do both steps 1 and 2 until the calculator cannot register any more numbers. (This leaves all particles open to evaluation by dividing the boundary.)

4. Similarities in numbers show unity.
5. Duplication ratios match with equal and equal opposite numbers.

Ratio

1. The ratio is a decimal point.

$$2 \longleftarrow \bullet \longrightarrow 1$$

2. Every number to the left, the outside influencing energy (sociology).
3. All numbers to the right of the decimal represent the energies that "keep to themselves" (psychology).

Equations

(Continuing Equations)

1. An atom is a cube, equated as c^3, captured negative space within the velocity of light.
2. Attractions of two atoms is equated as $2c^3$, the sharing of two cubed masses.

$$G = c^3$$

Within all captured mass, gravity is found. The outlined principles of energy replicated ponders factual electromagnetivity (positive and negative charges inside the atom). And the occurring attention of its mystery is being conducted through particle colliders, huge devices colliding particles for research. The research is attempted to link the secrets of the atom with gravity. The cubed equation of c^3 contains the same equisitiveness. The $G = c^3$ (8 15) equation indirectly obeys math already in existence for its own resolution of mystery.

Purity

In each mass there is energy. This energy ranges from the largest, the universe, to the smallest (which is the purest and easiest to calculate). From this, the unyielding quantity of flawless physics also overlooks its own miscalculations. The purest energy could abdicate its occupancy.

The purest energy in math is to take the smallest mass, the quark (q), multiplied by the velocity of light squared. The equation follows $E = mc^2$ for its simplicity. The purity exemplifies its true disposition. It is the true essence of life and nature. Through the purity of $E = mc^2$, we can keep it that way.

Numeracies

Man's numerical self is the guideline of life as we know it. Without these figures, it becomes impossible to see how genetics and DNA/RNA duplicate for reproduction and birth. The time and place when the pattern divides and multiplies is of utmost importance – numeracies attempt to define these patterns.

Everything has a destiny between time and space outlined by the very existence who created us. To pursue it may take another approach. This approach is a fine random assessment by using the theory of energy known as destiny.

Cancellation can distribute a higher foundation. As the future reveals more discoveries toward physiology, chemistry, biology, and physics, it need be said the pursuit agrees with actual truth.

Proof

The quark can be proven in its own right by relating indirect extremes. Doing so requires that the size and mass of the quark (size and mass being equal due to its genesis) be multiplied by the velocity of light – then square rooted. The answer to this relationship is the opposite.

<u>Example</u>

(proof)

q × c, then square root = 0.00256

Also, the quark equation does represent another negative-space difference. When the quark is multiplied by the velocity of light, it yields number 66, which is negative space of both radiuses. This also shows that the model theory 8, 4, 2 could work consistently. "Everything is equal to itself in a 2:1 ratio" becomes a part of numerical canceling. The ratio is equal to the quark, again multiplied by the highest mass squared. This represents the boundary and its outside.

Quark Velocity Spin

Each element contains particles and subatomic units. These units are the electron, proton, and neutron. Elaborate calculations can show resolutions to the rate they orbit and spin. These are velocities of each subatomic unit and quark.

Velocity

8 quarks per electron
a. Electron – 25,000 mps (q × 25,000 = 0.0000008)
b. Quarks – 3,125 mps (25,000 ÷ 8)
c. Unit – 25,000 mps × 8 quarks

4 quarks per proton
a. Proton – 12,500 mps (q × 12,500 = 0.0000004)
b. Quarks – 3,125 mps (12,500 ÷ 4)
c. Unit – 12,500 mps × 4 quarks

2 quarks per neutron
a. Neutron – 5,000 mps (q × 5,000 = 0.0000002)
b. Quarks – 2,500 mps (5,000 ÷ 2)
c. Unit – 5,000 mps × 2 quarks

Outer units of electrons orbit, while other units use centrifugal force to keep the balance mastered. The base is the velocity of light by each standard.

Sequence Differ

Changes in rates or standards of same numbers display mechanics. The overall unification may be construed as the same principle. The same numbers in different orders also fluctuate according to their functions. If the functions do differ but contain the same masses, energies, etc., this is sequence differ.

These differs multiply and fascinate substandards, standards, and evens. The proximity of these occurrences rectify how they correlate with monopolies of abundance. The prime example would be to travel the same paper route in differing times, streets, or roads. The job gets done, and the point becomes well made.

A case to reckon its versatility would be to measure physical unity. The examples are as follows:

1. Ceasing orbits or spins
2. Rearranged orders of function or shuffles
3. Birth of mass
4. Parallel equivalents

Time Travel

The inevitable question that surrounds the world of physics is the degree of time travel. How is it possible? Where does it come from? Why did it reveal itself? All of these are dominant questions so truly searched for.

If time travel is possible, it would be incredible to look into the future. It would be awesome to go back ten years ago to do the things you never had a chance to do. The possibilities are endless, but only one thing stands in the way.

Paradoxes are flaws in time where, if someone were to disrupt the destiny, changes would take place and people would vanish. If a paradox did occur, thought of theoretical earthquakes twice the size of the largest quakes would shake the earth.

Something as physics calls for a brave new world of engineering and medical care. If time travel becomes reality, we could look into the future and find genetic codes, the unified field law, and the answers we need. If sought after, it could become the great final human destiny.

The equation for time travel is of ceasing orbits. Once the orbit or spin of mass ceases, everything else around it travels in time, equal to quantum physics and relativity. The quantum mechanics suggest that while the mass remains at a standstill, the pull of gravity causes certain masses to collapse.

The time travel equation is an example of the foundation that gives us life, the quark. The new symbol for the time travel equation is m because time travel skips the time line that everything follows. To derive *m*, one must follow this sequence: 256 squared, divided by *c*.

1. Take the velocity of light, c (200,000 mps).
2. Divide the velocity of light by infinite mass energy (4 10).
3. Multiply the infinite mass energy by the highest mass squared (256^2).
4. The number 6.5536-04 represents the proton highest mass squared before the quark ceases, then collapses.
5. The ceased mass of the quark becomes enlightened due to gravitational force fields not having to use the same amount of energy.

Enlightened Mass

"When mass ceases, it becomes lighter."
The energy required to keep mass at a standstill requires less gravitational forces than when moving. The overall weight of an object is measured by how much energy it requires to keep mass complete.
When mass ceases, it becomes lighter. Physics law states that all forms of unity require some type of opposition. Thus concluding that less is needed when things do not weigh as much.

Example

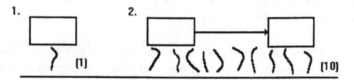

1. Object 1 needs level 1.
2. Object 2 needs level 10 while moving.
3. Both examples show the "enlightened mass theory" – when mass ceases . . .

To divide the velocity of light is to open up new avenues. The philosophy is the black hole, while containing the orbit or spin, ceases time travel. The black hole is a predicted star collapse, and through indirect physics, it becomes reality.

Another "time skip" example is to measure distance. Things farther away do travel faster. Take a yardstick and measure with a pivot near one end and move it slightly. The one end near the pivot only travels a small distance, while the other side travels a great distance in the same rate of time.

Example

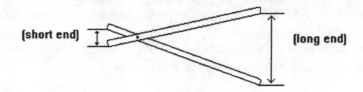

When compared to the "short end," the "long end" traveled much faster. Sometimes, simple examples provide statistical equivalents to be measured later, even if the example has already been resolved. There are always more ways than one to resolve ideas. That idea concludes that furthering education can be implied to enhance stimulation and attract students to be constructive.

Fusion

Understanding time travel puts us in a comfort zone knowing that certain masses can be disrupted enough to cause change in its natural pattern sequence. The deviants to fuse two atoms together for a million and some degrees of heat would be an amazing feat. The positive attraction of charged protons within the nucleus sets the composite for the fusion equation.

Physics is not understood by most, but the greatest thing of all is the written word. Written communication brings this world together. It enables us to solidify our thoughts. This is good for barriers of mind. It enables us thoughts of positive attitudes. (The negative space found in cubes separates the positive and the negative for a purposeful reason. This reason is balance and harmony.)

Energy can be useful to pathologies of the energy foundation (EF). The sun that bursts into millions of degrees is our example. It is the center of focus that keeps our solar system intact. It is gravity of the highest. The sun is surrounded by the highest mass in our high reservations of negative space (or ∞v) and positive energy.

To bring everything together in this world is human fusion. To do so through elements, however, two elements of hydrogen and helium must unite. Hydrogen contains the atomic number 1, and helium is 2. If combined through both infinite masses, then the number equals 3. To double the infinite energy would be a range of energy equal to 4 30. This highest range of gravity is the element that could be the renaissance to warmth.

<u>Fusion Equation</u>

$EF = 12{,}500^2 \times 16 = y^{x\,3} = \times 256$
(Punch in these numbers in exact sequence as specified.)

1. 12,500 proton velocity spin

2. 16 personalities – boundary/negative space

3. $Y^{x\,3}$ cube – infinite mass

4. 256 highest mass

<u>Gravity</u>

$G = 12{,}500^2 \times 16 = y^{x\,3}$
(1.5625 28)

58 squared four times to directly equal gravity (1.64 28).
When mass comes together with electromagnetivity, unity exists.

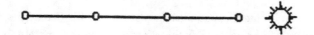

Closest Proof

In equations so surreal to the mind and math, we rely on synthesis and stimulation to prove that our Creator does exist. Even if the idea appears to be hypothetically distraught, it does arouse curiosity.

Creative math is the closest proof to biochemical standards we all contain. Numeracies dilate syntheses and extremities to prove that something outside of ourselves exists, perhaps our Creator. The unrelenting intuition deploys an appropriate just cause for its own endeavor.

Psychological physics is the indirect pathology not reluctant but an overview of drastic proportions. In judging these ideas, one might think that the creator need be locked up – not to disrupt the flow of things. But in understanding that physics is crazy in itself – traveling satellites bringing back images one thousand light-years away – the shock becomes lessened.

The closest proof to existing creation is to multiply the size and mass of the quark by infinite mass energy, 4 10. Infinite mass energy is to one existence because to have the velocity of light, which is an energy in itself, mass must be present.

<u>Example</u>

$$q \times E$$
$$(1.3107\ 00)$$

Squaring

The main man himself, Einstein, stated, "Imagination is more important than knowledge." The term *square* can be versatile in many regards. In athletics, such as basketball, to square is to stop or cease and make sure your shoulders and feat are parallel with the basket. If not, the shot will lose accuracy – unless skill plays a role. Accuracy is the deciding factor between winning and losing. Everyone wants to win.

Winning is recognition and the ability to place oneself above others. Usually, the recognition of rewards comes after the effort to excel is complete. And once the goal is reached, the right to receive the reward or award becomes justified. For example, you would rightfully receive a degree once you complete the curriculum. The often-overlooked sequence differ is literature and math – but the greatest sequence differ of all is science. The degree represented for each of the three fields shows that the development in expanding your mind has been accepted. So to cease expansion is to square.

Indirect relationships unite for teamwork and commitment to excellence. Pride also causes oppositions to cancel each other when the game gets underway. All energies of both positive and negative conflict throughout this disruption. But when the game ceases, both interact and shake hands, otherwise interact as they shuffle from one person to the next. In physics terms – gravitate. It is sportsmanlike to do so. The mind tells us shaking hands gives us a "truce of fight."

"The energy of man is his mass multiplied by the velocity in which he squares his ideas."

To square

1. To become parallel with
2. To collide
3. To stop or cease
4. Birth of mass
5. To cancel
6. To measure
7. To create
8. To capture

Big Bang Equation

Physics has an overall theory about the universe. The universe started which could hold the from infinite heat, or infinite energy, to begin creation. The speculation is, there may have been a beginning; but by general common law stated by common sense, "for every beginning, there is an end." Understanding common law puts us at a wall; we do understand this principle but do not have the knowledge to base a decision. There are limits to what we can explain at present (*Scientific American*, October 1994, 44). This being the case, there are questions that need to be answered.

Slowly but surely, the massive computer codes are plentiful mysteries to define high energy. The extreme application of power hinders faults due to common error. Inside ourselves holds the resolution that foresees common man. The gratuity of life is to search for answers.

When we seem to hit a void without further stimulation, the fight for the truth is a disgraceful boredom. We cannot continually estimate material without calculations or new theories. The big bang will eventually be explained, but not without other ideas. If we could find one, we can create creative insight to compensate for the lapsed time between theory and resolution.

The big bang equation pursues the new. Using the unified-field interpretation that every energy has a bond or function with other quantities of high resolution, it becomes a realistic viewpoint.

Although the examples are of a different nature, it does question such great forces of creation, such as the big bang. The explosion that gave us the planet upon which we breathe, is the final note to present gaps justified elsewhere. Values initialize great leaps and bounds. These values of quantum do justify those quanta.

When the universe is put at an athletic forte – galaxies against galaxies, solar systems against solar systems, particles colliding, black holes pulling, suns bursting, comets orbiting, and man calculating – we can begin to overlook the battle for superiority as spectators. We as spectators, comets as the officials, and the universe as the game, who will win? It is the grandest athletic battle ever devised, and we are waiting for the conclusion. ESPN is displaying the highlights, the Discovery Channel is explaining them, while *Scientific American* officiates its worth. They are squaring the ideas known as universal legends.

As man created the square, he, too, has to use it according to the laws of nature. Planets are orbiting and are trying to find their destiny. Black holes are capturing anything that comes their way. We are watching everything. And the time has come to be the official. The laws that underline creation become their own. These are the versatilities.

As explained through athletics, once interaction between teams stop or square, they come together in a joint – commune. This influences others to do the same. It is law of man to do so; therefore, masses measured creative cubes can be expressed. All mass holds electromagnetivity found to equal gravity, measured in small scales, whether being molecules or chemicals, the attraction of masses that give us the interaction of all life forms, otherwise known as the attractions between everything.

The big bang can also be measured or studied on small scales to equal its monstrous proportions. But first, the smallest unit, element, must be declared as a standard unit, c^3. Once a standard unit of measurement is set, the orbit or spin, found as ϖ, can be rendered.

If, for example, the orbit or spin of the object ceases, the opposing synthesis would not to hold together. The mass would collapse. Henceforth, the big bang equation.

<u>Big Bang</u>

Big Bang $= go^2$

1. g equals Bogartium, garth; 0.00000016.
2. o equals orbit or spin; pi, ϖ.

The sequence differ of comparing the cube versus the creation of the universe could be the prelude the resolution of the unified field. Everything changes in patterns – this is the explanation of the creative math.

1. Garth is 0.00000016.
2. Orbit or spin as pi, 3.1415927.
3. When multiplied to the squaring of the orbit, the number is 0.0000158.

These numbers are the sequence as followed by the trilateral. The idea is the orbits will stop. This would cause everything to ricochet under Infinite Heat. For every beginning, there is an end. The beginning would be the end. Everything would start over.

Prediction

As the universe collapses, the mass becomes so high,
it condenses into collisions, creating a universal size
equal to its collapsed mass size (black hole), which
transcribes a portal in time for a new universe.
The universe recreates within another destiny.

Big bang is equal to time travel.
Time travel is equal to fusion.

Immeasurable Quantum Time

$q \times \varpi$, equaled, then square root (0.0000101).

In every math form, it is imperative that the sequence be explained in full.
It is never known who will eventually benefit science. Anyone can benefit
science, whether or not they understand the formulated patterns.

Neutron: nonmechanical subtlety (101)

Trilateral

(Number Equivalents and Further Explanations)

0 – immeasurable time

1 – boundary or gravity (the downward/upward pull, ↑↓)

2 – to square, the attraction of the strongest force on the shape of the quark. the quark count of the neutron

3 – charges per nucleus (one-dimensional side of quark)

4 – proton (hemispheres of the mind, square of quark #^{2}, strongest side and highest surface area)

5 – completed forms of mass (planets, subatomic units, etc.)

6 – measurable negative space, quark count of light (neutral charge), pressure inside the atom

7 – electromagnetic disturbance or interference (earthquake, storm, etc.)

8 – electromagnetivity-tism

9 – energy (extended electromagnetivity past the atom's orbit of electrons)

In the number six (6), it is easy to point out that the charge for light is neutral. All neutrals are free to roam and do attract to other mass. Their neutral behavior allows them to be free from relativity (except for their own rules) but remain neutral to all other energies.

The equation $E = mc^2$ can display why the velocity of light was miscalculated. When two particles collide, energy is lost or gained. These losses or gains are slight, but relevant. The energy-loss equation is as follows:

<u>Time Travel</u>

(Quantum Energy Loss)

$$\cancel{E} = \underline{m}$$

Energy is neither lost nor gained as an infinite existence.
When energy is lost, where does it go?
Energy loss equals m.

When energy is lost, such as the energy-loss equation suggests, which is 66^2, time and energy become immeasurable within negative space known as ∞v. When loss of energy occurs, it cannot be measured. A zero is expressed for loss that is not obtainable by any measurement; thus 66^2 is the representation of the velocity of light.

The velocity of light miscalculation follows Albert Einstein's creation. Six is the quark count of light. The collision upon an object is the mass of six, which is negative space. To multiply the negative space by the velocity of light is to leave the number squared.

Miscalculation

$E = 6\text{-}\infty V \times 200,000$ mps (given) or 6-quark count of light squared
In other words, the velocity of light is a given because mass is equal to the velocity.

$$mc^2 = 66^2$$

1. $432^2 = 186,624$
2. Calculated velocity $= 186,282$
3. Differential ratio $= 342$
4. Highest mass $= 256$
5. Sixty-six squared (66^2) $= 4356$ (to square is to cancel)
6. The velocity of light is an even number, even constant.

Trilateral Configuration

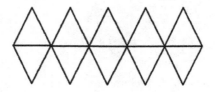

1. There are twenty-five shapes – five across, five down.
2. The final number to equal the Planck scale is the paradigm.
3. Point of connection: the point where all energies of the element studied correspond with another element. The point of connection are numbers such as 0.0000001.

4. Other points include each angle in the paradigm. Each represents an element listed, starting counter clockwise for each triangle (standard function).

5. The paradigm will display the unification when realized that all similar numbers have meaning of unity. If worked correctly, it will display the unification of molecules and compounds, or proteins for genetics.

6. The most important rule of all is to realize that certain energies change, or fluctuate. Keep working with the paradigm and the Planck scale until the proper unification is found.

7. Reminder: all things are equal to itself in a 2:1 ratio. Look for it; abide by it.

8. The trilateral also represents the dispersal of energy upon the completion of atoms, otherwise known as the strong force, which keeps atoms complete.

9. When two particles collide at the velocity of light, they leave a trace of energy.

10. The pattern for the complete unification is derived from $E = mc^2$.

11. The decimal point is the point of connection.

12. The thorough clues of the trilateral.

13. The point of connection is the birth of mass, conception. The immeasurable time indicates that there was another energy between the point in time of conception until its completion.

Strong Force

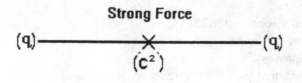

$$(q) \underline{\hspace{4cm}} \times \underline{\hspace{4cm}} (q)$$
$$(c^2)$$

14. There are five sides to the quark equal in shape to a pyramid.

<u>Equation</u>

0.0000066 reciprocated (1515.151515)

a. The square surface area is the strongest side, which is the beginning of attraction to the strong force.

b. The attractions of positive and negative are outlined by the direction the quark is facing: positive-outward, negative-inward.
   Arrows: the proton keeps the electron within its boundaries.

15. All squaring are sides shared at time and place (see explanation below).
16. Nonmechanical subtlety: element divided by element.
17. The final rule is to make your own rules.

Explanation

The point of connection exacts the force of the element. These forces correlate forces of energies and how they affect other forms of foundation: people, gravity, land, mountain ranges, the weather, chemistry, biology, and all sciences, etc.

In the model theory 8, 4, 2, the subatomic particles are the same size, shape, and weight divided by the placement in the atom. The mass of each particle is equal to the electromagnetic pull or gravity each particle contains due to its positioning. The only change we can discover are the unified forces that surround each measurement: the weather, the place studied, atmosphere(s) of physiological and environmental energy – all energies. All standards remain standards, just as though the number 3 will always be 3. It will be a challenge for us to find energies that are not necessarily detectable. The unified field states, "All energies interact with one another in some way shape or form."

Law

To declare the model theory 8, 4, 2, law, an equation must conform with all other theories of significance. The model 8, 4, 2 contains infinite mass within its cubed self. This equation is the proton velocity spin multiplied by the highest mass. Also, the unified field is equal to 12345678 squared twice for the 2:1 detectable ratio. This enables all forces to equal the force known as the universe.

All forces are outlined by what exact forces declare how it is discovered in a matrix. These forces are for the prediction of the trilateral. The pattern of the paradigm is defined by all great mathematicians: Newton, Faraday, Einstein. This leaves room for added curiosity due to great men devoting their lives to science.

The trilateral unifies nonmechanical subtlety. It may seem simple, but it has a series of similar sides. The idea is that all similarities are relative because all numbers ensue symmetry of the same energy source.

The element builds from collision of particles traveling at the velocity of light or $E = mc^2$, while the rotational spin indicates how many sides are shared at the rate of expansion before it becomes complete. The number of sides shared are equal to the degree of equilibrium in the infinite mass equations. These equations represent when sides of particles become squared, or parallel to each other.

Quark Per Quark

As opposing energies cross within each atom, they attract equally when the similarities of sides unify. These flush attractions keep the mass intact as a complete unit – and for the split second that they are opposed, the sides of the quark spin to provide formulations. In other words, once particles are parallel, the spin attracts other sides for its completeness. This is known as the strong force – force before gravity.

The twenty-six dimensions of equilibrium equal the inhabited unification of the Planck scale. When followed, the microrealm of negative space becomes reality. The ties, known as superstrings (direction of quarks and their attractions), are thought to be within the same realm as physical phenomena through the time travel and infinite mass equations.

Superstrings

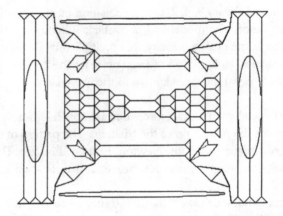

Once one side is shared, the equal opposite attractions of positive-negative charges cancel for other interactions around or within its newly created boundary. When two sides share, the neutron associates as a unit ($q \times 69^2$). As the three sides share, the proton begins its own venture to be a part of particle physics ($q \times 96^2$). After the three share, the fourth particle is attracted to display the charges in the proton. The fifth side of attraction completes a unit of mass equal to the number of sides the quark has. This then inhibits the distance between particles. The negative space is the junction before a disturbance of seven sides are shared. This interference enables for the eight-sided equilibrium to complete electromagnetism and the electron's charges. The final stage of equality is the detectable force of energy. It is only predominant when forces outside of itself display or extend past its own boundary. (The energy of nine is the same energy that opposes the negative space to impose opposite rotational spins that the atom has, the opposition is the opposite $- q \times 131^2$.) Once nine sides are shared, the ten shared sides skip measurable time. This enables us to visually recognize quantum time travel (quantum mechanics – relativity) within elements.

Zeroes are here to separate standard numbers, 1-9. Zeroes themselves cannot be measured. Everything travels in time, separated by immeasurable destiny of time and place – history.

These same energies of numbers cause fluctuations of force known as gravity. This force of the atom emits pulses or waves from its own moving mass. It also occurs at such a high rate that it appears as a constant – not a variable. The waves are also known as frequencies that have already been detected in numerous ways.

<u>Gravitational Waves</u>
(Attractions of Quotes)

"A line is an attraction."
"A frequency is an attraction."
"A line is an attraction to a frequency."
"A line is a frequency."
"A frequency is a line."
"A frequency is an attraction to a line."
"A line is frequent attraction."
"A frequency is a frequent line."
"A frequent line is an attraction."
"A frequent frequency is an attraction."
"A frequency is a line attraction."
"A line is a frequent attraction."
"An attraction is a line."
"An attraction is a frequency."
"An attraction to a frequency is a line."
"An attraction to a line is a frequency."
"A frequency is a movement."
"A movement is a frequency."

"All movements attract lines."
"All movements attract frequencies."
"All lines attract movements."
"All frequencies attract movements."
"All lines attract frequencies."
"All frequencies attract lines."
"All lines attract frequent movements."
"All frequencies attract frequent lines."
"All frequent movements attract frequent lines lines."
"All frequent movements attract frequencies."
"All frequent lines attract frequencies."
"All frequent lines attract lines."
"All frequencies attract eachother."
"All lines attract eachother."
"All movements attract eachother."
"All attractions cause movement."
"All attractions attract eachother."
"All attractions are movements."
"All movements are attractions."
"A movement is a movement."
"All movements are movements."
"A line is a line."
"All lines are lines."
"A frequency is a frequency."
"All frequencies are frequencies."
"Attractions are movements known as gravitational waves."
"Gravity is a movement of energy."
"Gravity attracts all energies of all forms."
"Energy attracts energy in all shapes."
"Gravity emits energy as waves."
"Planets are moving objects that emit gravity."
"Moving objects emit gravitational waves."
Topography

<u>Energy</u>

$O = \triangle \square^2$

O - atom
$\triangle$ - quark
$\square^2$ - to square

Energy Transversal Theory

In the final preparation for the energy transversal theory, it is important to pinpoint some creations and where they come from. Here are some predictions:

"If the brain created it, then it must exist, divided by its coherency."

"A calculator holds the universe within because the universe created it."

"Dates and time are equal to what everyone is thinking at the moment, divided by their personality and subconscious."

(If you divide the subconscious and personality of every person, then there would be only one unit or level of understanding, time itself.)

"All numbers come from biological states of mind, or BSM, which is our capability to understand the present physical matter."

"The body knows every secret in itself."

(From the point of conception, every energy has transcribed, locked away.)

"All numbers and what we see are relative."

"The human does contain the capability to understand his destiny."

"Only time will tell."

Conclusion

The overview is not necessarily reluctant but indirect to all proportions. Creative math follows new rules of its own for its own purpose but does not hold itself from any other math. This creates another proportion bound by Psychological Physics: Mind Before Math to influence the extreme of creation. This characterizes mathematics at its ultimate simplicity: creativity.

BOGART

Structural Psychology

Terms and Definitions

Myers-Briggs Type Indicator

THE MOST PROFOUND idea to life is death, and to ignore it is denial. AIDS is being denied every day. People are having unprotected sex, killing themselves before they are at the age of thirty. The rate of people projected to have HIV, a form of AIDS, by the year 2000 will be 40 million, from 14 million in 1994 (*And the Band Played On*, HBO, 1993). If that projected rate keeps up, it will be of consequence to believe we have a major problem because eventually it will look at us face to face. When it does, we will not be able to turn away.

The search for the solution is a bombardment. Science is spending billions of dollars funded by the government. The spending has grown from $23 million in 1983 to $1.3 billion in 1993. For 1995, the National Institutes of Health (NIH) alone has requested $1.4 billion (*Rolling Stone*). From that, it is understood that AIDS is highly difficult to defeat. The efforts to halt ferociousness lead into chemistry and high-cost drug manufacturing. The manufacturing is a science to keep certain guidelines and to follow them accordingly, thus keeping things controlled. Since turning toward complicated research to benefit overall production, the matter has worsened.

It is important to note that shortcuts and time savers within creativity could be another possibility not extended into. There are two halves of society, the perceptive and the judgmental. The judgmental are the decisive and controlling individuals that typically keep creativity, as an accent to life, subordinate to the greater, whereas the perceptive are creative and sporadic and keep controlled decisiveness opposite to the judgmental.

These differences tend to label each other and cause disruptions where understanding should be understood by everyone. The perceptive individuals are the most creative, while the judgmental keep order in place. Sometimes the judgmental are labeled by the perceptive as being "half there," while the perceptive types are labeled by the judgmental as "wandering aimlessly." But let us not jump to conclusions; instead, use what we have to administer the probability of mankind. The idea is to use perceptive individuals to create ideas, while the judgmental find order in them. This can give us diverse clues and approaches either directly or indirectly related to the problem at hand. Simplified scientific literature is the proposal. It has a lot of things science does not, the physical extreme using fight or flight. When one feels in jeopardy, he or she will make a quick choice whether to stay and fight or leave immediately. But what if the person cannot escape the danger?

We are all in the same boat of extremes. These extremes are diseases and viruses with no known cure. AIDS is the prime example of human danger. The threat will not stop until man outwits it. The time has come to do so. All precautionary measures can defeat what lives because what lives will eventually die. We only need to speed up the process.

Motivation

The basis for SSL is to use what we have. All forms of existence have meaning. The voids may bridge the gap between the time of resolution and lives saved. Creativity comes from the physical self. To put the physical self in jeopardy is to ensue great measures of ideas to battle and push away the fear. From a threat comes the best ideas to emerge from for studies of the physical self-physiology, chemistry, biology, and physics (Mead, Creativity Theory, SSL, 71). The theory is that our physical nature can automatically fill the void if pursued correctly. If the creativity theory is correct, the threat of dying could enhance solutions. This could be, in fact, the best new idea to come about in a long time.

SSL is a note-taking literature that responds to spontaneity, wishful thinking, and goals under extreme situations. The written pattern is to put ideas into short, conclusive quotes concerning whatever happens to be on your mind. The creative physical gesture is simplified into its true essence. The influence is then put into effect for others to derive their very own ideas and conclusions for directed opportunity. Even if SSL is not accepted by a wide scale of people, possibilities for personal motivation can be adventuresome. The advantage of these tools can enlighten scarce uplifting attitudes in patients themselves. Hope is uncommon when one knows that they are going to pass away before their time. So would it be possible, in listening to their fears and ideas of any sort, and be objective, but not face-to-face, for capturing sensational physical clues of fight or flight?

Perspective: if we use references for libraries, why could we not use notes as references? There will be a proper format devised as the idea aware to all, not necessarily in the format I use. I am sure the first caveman who invented the stick figure was quite wacko. The result: Pablo Picasso.

Attractions

While meddling around with spontaneity, man can put his best foot forward and declare his own destiny. Attractions can be found everywhere and are typically the center of attention. An equation can be expressed to show exactly how things do and do not attract. The equation is $2c^3$, the sharing of two cubed masses, or velocities traveling at the speed of light. The equation to present its center of attention equivalents is to take the quark times the velocity of light doubled. This expresses the boundaries of the atom and the three charges in nucleus. The attractions of mass and their cancellations of molecules can be united through psychological physics and atomic charge.

Center Equation

$$q \times 2c = 0.0000131$$

Attraction of Mass

(Molecules and Chemical Rules)

1. Use psychological physics – atomic charge.
2. Always cancel same charge.
3. Each subatomic unit contains equal number of positives per negative.
4. Cancel nucleus.
5. Cancel each subatomic unit separately.
6. Each attraction for calculated arguable trace patterns.
7. Attractions are as follows (not necessarily in these orders; that is for science to figure):
   a. N^+/N
   b. $N/N-$

c. N^+/N^-

d. N/N

e. Can be used with or without neutrals.

8. Cancel the units or groups for elements in molecules, etc.,
as being one. Example: water (hydrogen, then oxygen).

9. Final rule: the nucleus attractions come first, then
the electrons declare whether or not it is a covalent or ionic bond.

10. If all else fails, find a combination.

Sodium [NaCl] – neutral

Calcium N Chloride 17

Communication

Attractions in electromagnetism are more abundant each day. Science is pursing phenomenal ideas. We are departing from the old for the new and are looking for the path. Math becomes an insatiable candy. And as we walk down the aisle, it becomes Pollyannaism to simplify.

The knowledge for keeping in touch is a ravishing environment. We can keep on current affairs through television, news, telephones, radio stations, and talk radios. The significance is our silhouette of human nature. There is an emancipation to lessen the distance between people.

Those who do not communicate through technology use magic. This communication is the most shared secret that is always kept to themselves. There is a particular nature about sense and this communication without words. It is a definition defined by subconscious reasoning.

If physics contains the potential for an individual to page another from a beach across the world or to fax your business any time during the day without direct lines, then sensing other people could be within the realm of the same energies established for its cause. Something does exist outside of our consciousness. The chemical attraction is something gravitational. Newton, one of the greatest mathematicians of all time, worked his whole life to bring mankind the laws that keep us on our planet; but no one, to this date, has taken physics into its own simplicity to justify attraction of people through the Myers-Briggs Type Indicator – to map out what people are thinking at a certain point in time. Frequencies are but the difference between us all.

Interaction

The mind is a very mysterious thing. It is articulate, persevering, mathematical, and, best yet, caring and compassionate. It is the device that decides the fine line between right and wrong. It is eccentric in itself. Life is controlled by it, adhered to it, destined to figure its own faults and/or optimism.

Remember Fern, the little girl in *Charlotte's Web* who begs her father to spare the runt of their sow's litter? The attraction came through understanding and putting her stereotypes and judgments behind. She stood up for her belief and saw value where others did not. She had the inherent ability for value, and that put her own occupational hazards in jeopardy. The protecting sacrifice developed the relationship. Love for one another uses this sacrifice. One has to wonder, Is it worth it? Of course. It is a cause-and-effect relationship, but what drives this conviction?

The calculating nature of the mind triggers certain symbols and numbers construed by our subconscious. These are writings that bridge the gap between people. Within these scratches come specialties of measures. The mind is weighing the ties between communication. These emotions are expressed and shared by each individual. These are actions of the mind to retain frequencies that ensure each couple that they are thinking about each other.

Personalities

Personalities are the differences that make us complete. Some are rare, while others are domineering in society. Each has their qualities. The composure of percentages provides flair and uniqueness, which categorize people into different groups based on preferences.

When dealing with AIDS, it is a safe assumption to believe that there is a pattern or cluster of personality types and those who have the virus. Is it possible to combine the two? If it were possible to combine the two (which is possible), science and research could study the chemistry of those exact types for ideas of resolution through education.

Educating people on their behavioral patterns can put itemized emphasis on resolutions. Through the "typing" of people, it is possible to see the conditioning as percentages change from the current listing. As time passes, the growth and expansion of society preserves as a biochemical function, not a conscious understanding capable of being changed. If people's natural thought pattern cannot be changed, we still can learn the differences that make us complete.

Listing

(Personality percentages consisting of both male and female)

E. .75 (both) I .25 (both)
S .75 (both) N .25 (both)
T .60 (male) .35 (female)
F .40 (male) .65 (female)
J .60 (both) P .40 (both)

Multiply each type to create the percentage.

To and Fro

Traveling around from gives us sharp images. These are pointed out in films, movies, and paintings. We are in a round-and-about way, the species going to and fro. It concludes that the past can be changed by substituting previous memories, and the future can be predicted by planning ahead. But to do so requires statistics and the changing of people's needs and wants. Differing personalities play a major role.

Synaptic Orders

The capabilities have been found to have limitations. NASA has sent telescopes into outer space to unleash the universe's hidden truths. One such telescope is the Hubble Space Telescope. It has brought back pictures 1,500 light-years away. And if light-years can be brought back to planet Earth, it would seem possible for calculating devices to adhere energies of the mind. Thus, the synaptic order equation.

There are two spectrums of the mind. One is the conscious developmental pattern; the other is the subconscious developmental pattern or dream pattern. The left half of the mind controls the physical functions; the right, emotional. Both cerebrums control consciousness, while both cerebellums control subconsciousness. The fine detail of both regions furnace certain peculiarities: right cerebrum, conscious emotional activity; right cerebellum, subconscious emotional physical activity; left cerebrum, conscious fundamental activity; left cerebellum, subconscious physical activity. The equation as shown.

Conscious Developmental Patterns

$$S_o = \sqrt{\frac{1}{256^2}}$$

Primary Patterns
Fundamental left: 1.5259-05
Emotional right: 3.9062-03
Subconscious left: 6.25-02
Subconscious right: 2.5-01
Safe Differential Boundary: 5-01

7.0711-01 (INFP)	9.7857-01 (ISFP)	9.9865-01 (ENFP)	9.9992-01 (ESFP)
8.4090-01 (INFJ)	9.8923-01 (ISFJ)	9.9932-01 (ENFJ)	9.9996-01 (ESFJ)
9.1700-01 (INTP)	9.9460-01 (ISTP)	9.9966-01 (ENTP)	9.9998-01 (ESTP)
9.5760-01 (INTJ)	9.9730-01 (ISTJ)	9.9983-01 (ENTJ)	9.9999-01 (ESTJ)

9.9999-01 (17th degree)

The 17th degree is when a person acts outside of themselves.

Subconscious Developmental Patterns

$$S_z = \sqrt[2]{256}$$

256 squared, square rooted
65536 = 6.5536 04 "Actual Active Retrospect"
256 = 2.56 02 "The Division Point"
16 = 1.6 01 "Sixteen Personalities"
4 = "Hemispheres to the mind"
2 = "Halves of the brain"

1.4142 (INFP)	1.0219 (ISFP)	1.0014 (ENFP)	1.0001 (ESFP)
1.1892 (INFJ)	1.0109 (ISFJ)	1.0007 (ENFJ)	1.0000 (ESFJ)
1.0905 (INTP)	1.0054 (ISTP)	1.0003 (ENTP)	1.0000 (ESTP)
1.0443 (INTJ)	1.0027 (ISTJ)	1.0002 (ENTJ)	1.0000 (ESTJ)

The developmental patterns represent both consciousness and subconsciousness. Each number is a level known as biological states of mind or BSM. Through these levels, ESP can be compiled. The overall view is the first digit; second, left cerebrum; third, right cerebrum; fourth, left cerebellum; fifth, right cerebellum; integers, number of occurrences.

Biological States of Mind

In the mind, there are safe boundaries, and there are other levels outside of them. The safe differential boundary is the level kept to oneself before physical activity. The number that best represents safety is five (5). To be totally safe, one must never act. The only way to put yourself on a hierarchical pedestal is to believe it mentally.

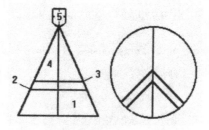

Biological States of Mind

    9 – Destructive level
    8 – Forever-life level
    7 – Highest physical level (sexual/comfort level)
    6 – Physical level
    5 – Hierarchy level
    4 – Conscious level
    3 – Retrieval level
    2 – Transfer level
    1 – Subconscious level
    0 – Simplicity level (purity level: something you cannot see)

Each is a nature of placement as being independent but still retains its overall function. All levels can occur in any region.

In the conscious developmental pattern or CDP, biological states of mind display the functions of primary patterns for each region of the mind. Taking one region at a time, the lower-left areas are the left cerebellums in each structure, and so on. Levels 2 and 3 are the division parallels between the cerebrum and the cerebellum. Each half of the mind accompanies these functions and works as a whole unit per half of the mind. When using the CDP equation with matched levels of BSM, each region fluctuates and gathers between the two. The overall level of regions is the first number (followed by the levels of retrieval and jumps), while the final numbers (integers) how many times the entire pattern must occur before the overall level registers as level one. If a zero occurs for any hemisphere, there is a short pause as it struggles to find new avenues or in the next phase of development.

The overall function is important, but not the most important. In reviewing the primary patterns of CDP, fundamentals are at the subconscious level, emotions are at the retrieval level, physical fundamentals are at the physical level, and the physical emotions are at the transfer level – being retrieved by the emotional conscious. At the same time, each hemisphere is represented by each proceeding number.

After primary patterns come the secondary functions, conscious developmental patterns. They have the complete itemized constraints, as well as ambitions. There are fourteen destructive and two differing personalities. One's overall view is of the highest physical level; the other is forever life. Each hemisphere of BSM shows their enthusiasm and what they prefer.

The foundation function of all humans is their subconscious. The subconscious feeds the reality of life, life itself. The subconscious developmental patterns, or SDP, are the hidden secrets below our comprehension and are equal to dream patterns, the thoughts of sleep.

What happens deep within our minds is the groundwork for dreams we pursue to keep stimulation present. The subconscious need not happen at all. It is not dependent on anything else, besides energy, because it does not know it exists.

Consciousness: everything depends on each other – the only difference is knowing it. In the dream patterns, the division level between both is a transfer level. The mass is the same standard before things reform at a higher level. The division point gives leeway for both worlds.

Indifference

The best example for rendering is the INFP. He or she contains the actual level needed to comprehend that the physical self is, in fact, the subconscious. For others, it is difficult to understand because their levels of the physical self are not at the subconscious level, at all. The truth is always indifferent, because each of us are. Also let it be pointed out that many people mistake the subconscious level, and besides, what else could it be?

ESP

Extrasensory perception is not just a fable or figure of speech. It does exist, but the perception of feeling energies may not be apparent to the individual. In the equation, the perception of understanding can put things into perspective. The rules follow the same pattern as both developmental patterns, typically shared at a subconscious level, then transferred to the.

The integers of all functions of the mind in creative math can be replaced by the energies of people or the number of people themselves. So, in other words, the integers can also be people's energies outside of the individual ($ESP = 2c^3$).

Things always appear to be easier when people involve themselves with others. The rate of exchange quickens, enhancement occurs. The mind becomes computerlike and sharpens its skills. This is the grounds for attending parties, social engagements, athletic games, etc. The mind senses other people's emotions and thoughts. When focused toward one individual, one can sense.

Sun

The center of attention calls for extremes when figuring how it reverses itself. When a person captures attention, they can feel it. The reversal equation uses the sun to show the idea of reverse ESP.

Many people have been told anything is possible: black holes, planets, galaxies, and comets. To show the relationship may be eccentric.

As it takes physical activity to outline the confines, it is important to search out and define it. The potency of figuring what other people are thinking is at its highest when all personalities are bound in a vacinity of one another (Observational Psychology, SSL, Mead, 74). When all energies are directed toward one cause, the purpose or engagement is understood. It engulfs the attention and becomes the highlight. If everyone were interviewed, the response would be ecstatic.

What draws attention to a cause? Is it what others think about? Do people need to express it for it to be understood? Can it be expressed and sensed physically? Yes, even without words. The levels must be extremely high, though. It cannot be shared until people have the same ideas.

Compiling thoughts for others to see requires paper. To come together is a form of squaring. All opposing energies or extrasensory perception come together within the mind, square themselves, equal it by writing it down. As whit and whim of idea gathering come into translation, it dwells around the rarest individuals who emphasize togetherness. These individuals need only one person to spark their subconscious to enliven it.

ESP reversal happens when the energy foundation squares itself for others to correlate in both directions of any voluntary physical activity. It must be seen sixty-one times, or noticed by that many people, for it to reach the individual. Once opened, the channels conclude unity of thought. It surrounds us with this idea in mind – frequencies are what they are called. They give us life. It is what everyone craves.

Once the individual does not have the attention within themselves, they are burnt out. The attention wears off – a fight occurs.

<u>ESP Reversal</u>

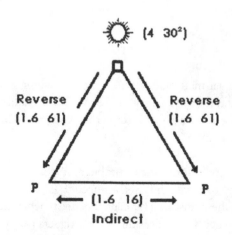

The sun is the energy that gives us life. Its warmth brings us together. And when it brings togetherness, we all have a need for sharing. Its influence sparks out minds and stimulation of curiosity toward one subject. This causes particular collectives upon something most admired. Once understood, it correlates with a pattern of people that gather for its magic. When this does happen, it becomes recognized. These are the dynamics.

Division Point

The division point is the most accepted of all energy sharing. It is the body and mind of both people. We are constantly transferring each other at fundamental hierarchies while being emotional. The integers pinpoint the precise numbers needed for it to occur. A wonderful thing takes place when each share. Birth is the most prized possession we have.

Engagements and a lifelong commitment is forever love of one another. Our ceremonies intertwine our companionship and trust. Love is fusion. The urge to share without words is our family. Our families model themselves around faithfulness toward an once-in-a-lifetime event. Marriage nests our seasons. It silhouettes our manner, creates inner peace. The need to express with others becomes astronomical. Attention is a must. Influence for feedback, reverse ESP.

It influences others after honeymoons. It has a tendency to overwhelm us when others do the same. The love is to give the couple the bond. Those who have influenced others to do the same, seventy-seven people must become aware. The energies have come back around ($4\ 30^2 \times 2c^3$).

Chemistry

AIDS equivalents equal math and the mind. Math within it is the true mystery dawned by science and man. When scanned by CAT scans and other instruments of high-energy physics, could be possible for certain energies to be lost. The probable error could cause skepticism.

Chemistry between people also causes a problem. A lack of it makes things stale. In order for relationships to remain intact, they need to be adjusted day by day, or the attention for one another perishes. This can influence biochemistry within the mind. Breakups could endanger the relationship. What triggers people to avoid truth?

Social science is trying to define the virus so the overall complexity of people can be avoided. Anyone can probably see why. Such instances frighten anyone. Since it is such a mysterious virus, we have given the problem to the science fields of chemistry, biology, and physiology. Could the inclusion of types help map out the pattern? If so, why not pursue it? We first must understand what we are looking for. This could give us a place to start.

Influential degree: the young, middle aged, and
the elderly are listed from easiest, respectively.

dominant – young

subdominant – middle aged

subordinate – elderly

The End

FIRST OFF, I would like to say that these creations I have brought for us to see are not as easy as they seem. Ideas like these take an extreme amount of time. It is not every day that someone can start off with the universe through physics and then bring it all the way back into psychology using a new math never seen before. But worth every last drop of energy I could muster. Long nights, no sleep, avoiding distractions, overcoming the ability to quit, the dream to be the best, and the goal to surpass the unsurpassable are all goals I see before me. Each one plays an important role. It is even possible to pop up out of the blue and surprise a few people. It is a difference we must understand. For what we are about to see are major leaps in psychology. It is an up-close view of indirect physical relationships. My plan is to be tactful and to the point. It is vital not to waste time; to conserve leaves room for more. And the idea is indirect physics.

As a reference for "ESP and AIDS equivalents," Bogart Structural Psychology, Myers-Briggs Type Indicator becomes a marvel for biochemical functions. There are many opposing forces all of us must face in life, and to understand oneself before approaching life would ultimately make things easier. Education will always allow the mind and body to overcome our own understanding. The idea is to direct visual stimulation toward remembering what makes us complete. This may be the best reference.

For each half of the mind, there are three functions. These functions show the rate for each region and parallel. These levels also display biochemical dominance in each region and parallel. As chemicals interact between the cerebellum and cerebrum, they give characteristics of dominant, subdominant, and subordinate levels. These are coordinated with black, gray, and white.

The greatest misunderstanding in life is that we all are alike. That is simply not the case. There are standards for everything, and the standard for people are the sixteen differing personalities. When one can recognize the differences, the dynamics of subtlety become more noticeable. (It should also be pointed out that these symbols show why some people have an adapt nature – and why people label others as being somewhat insane. It is natural to do so because there is one standard mind that attempts to justify us all.)

The states of mind are not controllable but are influenced by sociological aspects. Even such things as music or the using of certain senses can enhance the ability or alertness. In fields as science and medicine, stimulation might enhance judgments in surgery or private practice.

Standards

The mind is based as though it was meant to have a particular nature. It has certain qualities found in every walk of life. These certain natures create functions always separated by the person's approach on life. They divide the differences between us all. They predict the social influence (which could mean everything).

The sensory organs play an important role in finding actions. The most important thing that everyone should know is that each hemisphere has its own nature. These give us fascinating distinctions in physical, as well as emotional, expression. The left half of the mind controls physical fundamental activity; the right, the emotional.

The rates of physical and emotional activity are dominant, subdominant, and subordinate represented by the pigments of black, gray, and white. Each rate is the transfer of quickness from one region to another. These transfers are separated by divisions. And each parallel includes other stimuli in the midbrain and the brain stem locations. The two parallels are as follows: egotism parallel, the left division parallel that indicates the degree of physical expression and control; submission parallel, the right division parallel that indicates the point and time of learning, which is equal to certain synapses and transcription of emotional memory (*Scientific American*, June, 50–57).

If looked into further, each standard of mind does not exclude certain functions that include other stimuli. In my structured symbols, other sensory organs involve the opposing outside energies and how they influence. For example, there are three other influencing forces of subtlety that can be displayed for both parallels: visual cortex, hearing, and speech. These three can both indirectly and directly provide stimulus to each.

The pattern for the visual cortex as follows: eyes, visual thalamus, visual cortex, cerebellum, and amygdala. Once the eyes translate what they see, the visual thalamus relays messages to the visual cortex located directed above the region for physical muscle control. Once the cerebrum receives transmissions, it quickly feeds into the amygdala for sudden extension of muscle translation through the brain stem. This entire pattern is included as each parallel.

Another stimulus is speech. It is located on both frontal and temporal lobes. The expression is how talkative a person might be. The final stimulus is hearing. It also demonstrates the indirect influence in characteristics. Each three enlivens the entire perception of sense.

> Parallels – Brain stem: medulla, reticular formation, pons, midbrain,
> hypothalamus, thalamus, pituitary gland

Types

For each personality, there are four per person; and if broken down, the visual stimulus can be enhanced by each individual who chooses to link visual awareness through correlation of memory. If they do choose to do so, *Gifts Differing* can check and balance.

The visual recognition provides clues. These clues can be the decisive factor to enable people to understand themselves. The common decency of man is to understand oneself before others do, the purpose being that when one understands him or herself, it becomes more difficult to go outside of what one knows or want. This, in turn, could cause an educational overload of a persistent variety. This variety could ensue an inner peace for everyone to see.

In typing, eight make up the sixteen. Each has a functional contributor to our world. These distinctions are labeled into groups. These groups are for easier understanding. The rules: the first letter must be indicated by extrovert (E) or introvert (I). The second must be either sensing (S) or intuitive (N). The third, thinking (T) or feeling (F). The fourth and last must be judgmental (J) or perceptive (P).

Function

Everything has a function, based upon Freud's theory of the id, ego, superego. The symbol theory of the physical self is the id, which does not know the difference between itself and reality; the expression of physical control, the ego; the emotional acceptance and auxiliary process as a part of memory, the superego. They all work in sequence – starting from the emotional to the physical self. If there were to be arrows, they would look like this:

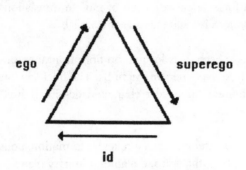

Male pattern: works in the direction above
Female pattern: works in the opposite direction

The peace symbol is different. Isabel Briggs Myers wrote a small inclusion on the back cover.

Distilled in these pages are the insights
gleaned during a lifetime of sensitive
and loving observation of people
and how their behavior reflects their
psychological type as measured by
the Myers-Briggs Type Indicator.
Myers believes passionately that
each of the 16 types has its own
strengths and that understanding and
using these can lead to fulfillment.

Differences

The extroverts (E) of the world prefer the outer world. Introverts (N) prefer the inner, their own thoughts. These differences are regions of the left half of the mind.

The sensing (S) individuals prefer a higher degree of physical control as opposed to the intuitives (N), who prefer the perception or curiosity of the physical self before reacting. The difference between extroverts and introverts is the egotism parallel.

The judgmental (J) prefer decisive emotional control as conscious emotion, which is opposite to the perceptive (P) individuals. Perception is the opposite to judgment. It is a physical emotion that tends to be more sporadic and does not follow the same form of control – a suppressed physical emotion, not depression.

Between the emotional regions, there is the submission parallel. This parallel is the auxiliary-process emotion – something more overwhelming than the ego. Thinking (T) parallel is emotionally domineering and has the ability to be quick. It is definitely more protective to what that person wants to learn, compared to the feeling (F) individuals, who have an openness in emotion. This region to them is very open to curiosity. They are emotionally open-minded.

If you were to put all of these together, you could see the tendencies for each type. To go even deeper, anyone's personality shows the characteristic for literature and science and where they come from. In the symbol theory, symbols and shapes come from the right cerebellum. These are to us as numbers are to a physicist, letters. The other cerebellum creates numbers and color. Together they create our educational means.

Education

Symbols often repeat themselves. The energy transversal theory is to define where these shapes come from. As we know, the best representations are the triangle (equal to the one-dimensional shape of the quark), the circle (cubed mass), and the square (highest circumference on the quark equal to mathematical squaring). Once understood that unity is a complicated definition, we must insist these shapes come from somewhere.

○ △ □

Dynamics

In physics, there is an attempt to resolve the unified field as a paradigm. This structural reference to unity comprises similar energies or forces to graph gravity. And if a feasible paradigm does come into existence, gravity may not be difficult to calculate. The deviants themselves could branch into new forms of creativity – divided by the creator.

It has been brought to my attention, through people's vision, that everyone has a higher remembrance rate once visual stimulus occurs. My attempt to define gravity through the mind may be correct because math came from somewhere.

The terms for each paradigm of nonmechanical subtlety for unity of all similar numbers at certain periods of time. Even Faraday has shown that electromagnetivity and gravity are of the same manifestations. The possibility to believe that a simple calculator could deviate gravity is not such a padded-room idea.

The listing of paradigms does follow the three basic shapes. They were built from my ego to build my own credibility in science – indirect physics. It is a probable cause, but eventually, everything is. To suffer the consequence of denial would be to deny its possibilities to enhance science in any way.

The created view of simplified scientific literature is that a simplified created quote can give influence or ideas to someone who has enough education to make a difference.

> When you buy the necessities for an immaculate dinner or meal, do they
> appear as they do when the meal or dinner is completed? No. The literature
> is only the spice to the overall flavor. Add your own. Each idea has
> sixteen interpretations.

In Depth

Once you find who you are, life could be easier. If, for example, one was confused with an abundance of information, certain repetitions of characteristics could be followed for peaceful resolutions. To understand oneself – nothing could be better.

In dire straits, we all have formed the healing of oneself. But if not put into probable physical perspective, the oversight might harm instead of heal.

The transfer of creativity comes from the division point. The transfer of physical action at fundamental conscious level is a hierarchy to place the act above everyone else, while at the same time, the emotional awareness level remains physical to place all physical actions above it all.

The ideas of the "inner child" come from ESP: first, the mind; and then from what the person sees in something else, which is the second individual. If one was to create the passing of a movement, the possibility to see the quickening of subconscious spontaneity becomes reality, even if the person has no creative blood or is not spontaneous. It can happen or be controlled by anyone, divided by the personality.

As I tend to repeat myself the warped style, differing sequence of twenty-six letters, it is imperative that I speak in tongue of my own. This is a direct link to people and how they see each event in society. It is a reminder that too much pressure of the mind foresees a collapse. All individuals know there must be some type of escape.

Laughter also plays a major role. The theory is that laughter is the inevitable hierarchy of both cerebellums. The cross-reference for both fluctuates the pattern for adrenaline and emotional suppression. But also, it is a physical relief to those who choose exercise as a replacement for all three. This gives reason for humor because it comes from the same "point."

Certainties

The mind cannot stay within itself without outside information. The sociological aspect can be harsh to an individual if not understood. Everyone becomes depressed, suppressed, or even angry in a lifetime. It is difficult to avoid, so we all need an escape because the mind cannot retreat when it desires. It no longer can capture the highest rate of sharing one can offer when too much emphasis is upon one individual. And as we all know by now, the simple spectrum is the prelude to the opposite. Einstein's simplicity of energy blew up Hiroshima. The power of the mind is a wonderful thing if used correctly, but it does have drawbacks.

In certainties, the dividing line is dominance. To study these characteristics, first, one must follow a certain path. The path is, in itself, one's own thought. These could mean everything, but greatness need not be mentioned, only in text. Literature is equal to speech indirectly; so therefore, if one was to mention each page and paragraph when speaking upon greatness, the person would sound like an idiot. Quite a few things in this world remain understood for a purpose. This purpose is for furthering education without always having to prove it because as long as one person knows the truth, there would be no point in lying so one could stand there looking dumbfounded when proven wrong. Knowing that contradiction is around the corner helps keep everyone in line.

"So if you do not know, keep it to yourself – someone may drive you crazy."

Psychology

In the form of structured symbols, the findings of observation suggest that there are times in this world when most people appear to be similar. The truth is the exact opposite. When we see the differences that make us complete, it becomes easier to relate physically, emotionally, and visually. And the idea of psychology is to study the state of the mind as it is.

Each characteristic provides help in governing how a person approaches life. They are repetitious, but so is life. Let us look deeper into our own selves. The poetic nature of life is the balance of certain releases each of us ensue. There are three balances in particular, and they are humor, the structures we build from dreams. Humor is a level of emotional dominance, structures are the level of physical dominance, dreams are visual physical transfers of dominant regions – cerebellums. Structural psychology defines these escapes of physical cycles.

Introversion

Introverts keep to themselves. Their inner life is more dominant than anything else. Their value in life is to keep control of physical expression or fundamental awareness. It is a function that keeps the outside world intact. Through the simplistic definition of symbol psychology, "introversion is physical fundamental awareness."

Extroversion

Extroverts love the outer world. Their personal life evolves around other people's physical nature, which includes everything. The extroverted value is the subconscious need for physical exertion at any level they choose. And as stated, "extroversion is physical awareness."

Intuition

Intuitives have curiosity. The need to find out what makes things tick is an unrelenting pursuit. They never give up and believe all avenues of improvement can break down barriers that might hold something from within. Their perception is quite different from sensing. It is also defined as "intuition is a perception of the physical self."

Sensing

Sensing individuals are extremely physical. There expression is dominant. They always have a need to be included in physical activity. It is as though sensing types have a quicker gift toward their nature. They even have the best physical judgments of ESP. "Sensing is a physical judgment of ESP." They can sense when one is stressed, happy, sad, etc.

Feeling

Feelers tend to understand the emotion within everyone around them. It is their destiny to make decisions based upon how they feel, not how the actual affairs state should be. Feeling individuals are constantly riding a wave of emotion up and down. It always amazes a thinker about how big an emotional roller coaster they follow. They are constantly reminding them to be decisive and emotionally controlled. "Feeling is emotional ESP."

Thinking

Thinkers are emotionally solid. Their dominant auxiliary process is to be emotionally quick and decisive. They are complete in all forms of emotional control. This allows them to withstand the hardest of times. The only drawback is that they must be reminded to have a heart. But on the lighter side, they are the best friends to have because they value completeness of family and friends. Their decision making is second to none. Do not ever cross a thinker's path. What you do not know might hurt you. They protect their emotion at all costs. "Thinking is protective ESP."

Judgmental

Judgmental types are controlling. Their nature is to organize and make worth out of cluttered information or events. The passionate desire to keep things organized often leaves them to believe that nothing has been overlooked, when in reality, the need the perceptive individuals to make sure things are not because certain facts still remain emotionally disorganized. Judgmental individuals automatically despise chaos unless notified first. They can often be called upon for the need of control in unorthodox situations. "Judgments are controlled emotional ESP."

Perception

Perceptive people are emotionally physical. It is a capability to understand judgments through someone's physical emotion. Perceptive types are also more complex (in most cases). Emotional awareness, as compared to the judgmental individuals, is an extreme. Perceptive individuals love chaos. Perceptive individuals need to be organized in situations that need it.

Both the perceptive and judgmental need each other just as everyone needs to be close to their opposite. It provides the harmonic balance in life. Perception provides the creative balance we all crave for. It is a perception of others' physical emotion. Otherwise known as "Perception is ESP of emotional physical awareness."

Awareness: Hemispheres/Regions
Stability: Parallels

Every one of us has their fine qualities. Not one is better than the other. The balance of people gives us peace. And to misunderstand this principle causes havoc or chaos. No one wants to see havoc. If we understand each other, this world would be an easier place in which we live.

Opposites

All opposites need each other to modify behavior and to take actions not typical of oneself. It is a reminder to each of us. Life becomes easier to pursue life through someone else's eyes. These are listings of the opposites of need.

N – S

"Intuitives need sensing types to be reminded that the opportunity may never present again."

"Sensing types need intuitives to be reminded that the plane for Pluto won't leave until the year 3000."

E – I

"Extroverts remind introverts to make physical actions."

"Introverts remind extroverts to settle down."

F – T

"Feelers need thinkers to be decisive."

"Thinkers need feelers to have a heart."

J – P

"The judgmental individuals need the perceptive individuals to be thorough before a conclusion is drawn."

"Perceptive individuals need the judgmental to be organized and on time."

Combination

We can begin to deviate pathologies of both psychology and sociology. Both can show just what functions expel unto the world, and which ones are kept to oneself. There is also a function that can go both ways. It is the easiest function to influence and only carries half completeness that can predict those influences.

Let us continue. All black characteristics are the dominant functions of 2. They are the influencing regions or parallels in society and within ourselves. All gray levels are the easily influenced. If one were to list the 2:1 ratio of characteristics, they would be as follows: black (2), gray (1/2), white (1).

In society, there is a tolerance level. It should not be crossed for everyone's sake. When two people conflict, there is a reason. Rates of function are equally represented as standards. Thomas Jefferson stated, "All men are created equal." Only divided by social influence. Within our society, the object would be the educational level of the individual.

If one was to take a philosophical view, a compatible theory could be devised. If someone was to meet their "match," a battle of brawn would occur. Same social levels of black will fight for supremacy because everyone likes to be on top: above and beyond all others, with the respect of those who challenged. This energy of fight causes an overwhelming conflict. It is probably the most overlooked. Some say it is constructive, others say it is not. Whatever the viewpoint, it is a confrontation.

When arguments do arrive, even if overlooked, one can see themselves set aflame when they take one step back. And if one does not understand the fine line, then it becomes crossed. Physical confrontations should be avoided but often occurs, anyway. It will always be a part of humanity. The fascinating question arises. Can it be traced, or is it an intractable problem? The answer is yes, no. Ratios are the deciding factors in fine-line assessment.

The ratio within everything gives us cause to believe that perceptions can outline influence for social violence and domestic abuse (SAD). When someone strives to better themselves, it is very difficult to reach that level above all others. The odds are not apologetic.

The pursuit for greatness is ferocious. The ego is hierarchy. All men are created equal. The need is perception of some sort. It is not a shame to admit that each of us wants to be invincible.

Balance

The new beginning for all of us to see is a personal war – my belief. We need to understand the differences between us all. These differences provide patterns irrelevant everyday life. All we see is what we know. It is the foundation of all memories.

There is a watchful eye of an angel looking over us all. We must follow her path and understand her speech. The speech is a pattern. Something created us, and it is important to understand. Our pattern of thought is psychology. The way we go about approaching is sociology.

The possibility of combining the two becomes reality through visual recognition. The negative space that divides us all is ourselves. The void of distance, which is negative space, can be filled. Personality type, for reasons unknown, have a division of the number 16. Sociology and psychology go hand in hand. This helps keep everything intact. To understand this principle is to enhance and enliven what all of us are looking for.

The introversion of the intuitive is a dominant function of the left cerebrum, a dominant trait in bringing new ideas for us to see (indirect ESP reversal – to square). Even literature is a form of ESP. It is indirect but nonetheless a transfer of thought that can carry the mind's deep convictions. The energy required to create anything is a transfer of directed energy to transcribe thought patterns, divided by personalities for metaphorical exemplifications. But for all of us to see it clearly, we must foresee that "no one has the privilege to know what is on the other side – until they get there." Let us not jump to conclusions before it is time.

Indirect Physical Psychology

Through the organization of judgments, we can render reasons why certain abuses or violence occur within society. The need is tremendous. From a talk-show view, it appears that people are fighting each other quite often. This poses a major problem for anyone or everyone accountable. A form of protection must be met for psychologists and psychiatrists to educate their clients so they can avoid their own faults. This could give us certain perspectives not previously recorded.

In indirect physical psychology, there are fallouts each of us have. These fallouts occur when we do not have escapes for aggression or anger. These aggressions could then be projected.

Love

The distance, which means "within the same vacinity," is decided by how "close" an individual is from one another. It is the most memorable experience one can have of one another. When people fall in love, it is a wonderful thing. Nothing, for a period of time, appears to shatter or disrupt the love tie. It is a relationship tied by adjustments agreed by both. But if tied too tight, anything could happen, even the seventeenth degree.

All of us have been pushed to that extreme a few times but should never be in a relationship as frontal assault. It is OK to walk away from those who you think deserve it, but only constructively. It has never been our nature to be angry at someone. As we all are, no one likes to be pushed around. As Tom Hanks said it best, "My momma told me never talk to strangers."

Breakdown

The final curses in indirect physical psychology are the energy shares we all have. Within the high-level share, it is necessary to govern the comprehensive breakdowns of how certain abuses occur. Through what we know about personalities, it is still important – I might add – to see all possibilities before conclusions are drawn. And these pathologies work everywhere, not just for evaluating abuse or confrontations – everyday life.

As stated, the mind is a standard. Examples of standards exemplify. These are those examples:

<u>Breakdowns as pathologies for ESP</u>

"It is always easier to influence something that already exists."

Sociology ●——→◉

black vs. white ●——→○
    2/3 influence
black vs. grey ●——→◉
    2/3 dominance
black vs. black ●——→●
    cancellation - fight for dominance
grey vs. black ◉——→●
    1/3 dominance
grey vs. white ◉——→○
    1/3 influence
grey vs. grey ◉——→◉
    complete function

Influential ratio

● **black - 2 to 1**
◉ **grey - 1/2**
○ **white - 1 to 2**

Stress

black vs. black
equals grey
●——→●

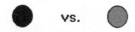

"Black and grey are the reasons for living, but they are also the dangers of life."
10/23/94

Indirect Physics

ESP and AIDS Equivalents

Introduction

THE FOUNDATION FOR all things has a beginning and an end. This is the final theory. It is my life. I will pursue it with care and skill because it requires that I do so. But it should be pointed out that all my ideas are new to this world. Even though they follow the Myers-Briggs Type Indicator, they are genuine.

Everything, including all acts, has a definition. These definitions are my ideas. Within these, I thee wed society and confrontations of any sort – social violence and domestic abuse – even war. It is a goal of mine to insist that these work. Nothing else makes any sense. The main goal is to find a channel to feed my ambitions. Use what already exists. Ideas always come from someplace else and have relevance not usually seen in the beginning, but toward the end of perception. This is my own perception of the world ($ESP = 2c^3$). Indirect physical series is equivalent to quantum mechanics, otherwise known as relativity.

Taking into consideration that there is a cluster that could combine medical problems and personality types, physiology could benefit with medicine through simplicity (Mead, Physiological Theory, SSL, 70). The statistics could later be directed toward each age group for further educational purposes.

"We could begin to influence young doctors to lend direction for our future generations."

Angst

Our life is the anticipation of future events. We cannot live without it. If we never foresee or predict, life becomes dull. No one wants that to happen, but it is a reality we all must face. For our own sake, we must believe in living forever. But with life as it is, it is impossible to avoid the fall we all have. This state of mind is depression. It zaps our energy like a great, big space creature that we often see in the movies. The battle is to beat it up before it gets our crew, our family – the human generation.

Cognitive theories of depression emphasize that when the alien of depression distorts our outlook on life, it changes our perception. It is nothing to be ashamed of because it happens. The idea is to chop off its head before the dark side overwhelms us. It is human to deny our mortality for our own sanity.

Changeover

As the mind goes through stages, it is almost certain that different levels justify its state. The mind, theoretically, developed math, and the change over follows the same idea. Each level that occurs is one higher than the previous level. And once level nine is reached, biological states of mind start over. In conclusion, the mind is equal to our standardized numerical system.

The conscious developmental pattern has shown that fourteen personalities have an overall destructive level, but when people attract, certain behavioral patterns change. When two destructive behaviors come together, otherwise known as ESP, the loop occurs.

"The AIDS crisis is equal to ESP – > go until we cannot pursue it any longer."

"The mind travels in time as a loop of thought."

$$9.8765$$
$$+9.1234$$

$$8.9999$$

Protect yourself: most personalities are destructive and when shared,
a belief in forever life.

Belief

Even belief is playing a major role. Since most people do believe for every existence there is a creator, it is easy to understand that certain people never regret risking their lives. It is still no excuse to jeopardize the lives of others. If psychological physics is put into perspective, one can logically assess similarities of math. Only one given energy source can create such events as the universe. Something that is crazy, however, is to avoid the truth. We all do it, but not all of us choose to deny the fulfillment of replacing it with knowledge.

There is a secret number kept to ourselves until now. First, one must understand that everything ever created through energy – everything we see – is open to interpretation.

The basis for the American flag.

Preachers are giving sermons and speeches. They are repeatedly reaching in the air to grab hold of something physically not existent. This feel we all have is known as an "uplifting experience," without necessarily the physical action (0) to show such. It appears that belief is a sharing of ESP as the closest energy to our spiritual leader.

What drives people to this enlightenment? Naturally, one must wonder if the person writing this is loony. The answer is no, just eccentric. There is also a trait. This trait is a belief in God, not necessarily the way other people view it. I have an independent judgment from all others, which is obvious from my own creativity. The best example people can relate to are two great men and one woman who devoted their lives to science: Darwin, Einstein, and Curie (exception: any extremely successful female). You see, all three had a creative ability that drove their spirit. We could always read what they wrote because they chose the written word. The point

I would like to make is that certain people are often overlooked because they follow a path all their own. This does not automatically give them an extreme amount of intelligence, but it gives them the frame of mind to pursue the things they like. No other person had to give them belief because that number is given when born. This number is the "closest proof."

What with, we can see the differences in sex.
– Male vs. Female

Nonmechanical subtlety is found in more places than one.
– Evolution

Conclusion

In conclusion, there are many things that we can do with ourselves: solve our own problems, create new avenues of improvement. You name it. What is an expectation? It is referred to as an ESP projection. When one does not know what to expect, it causes the mind havoc. It is not unlikely that performers cannot find the avenue of escape. Where can they go without being recognized? Once you reach the pedestal of no return, you have no choice but to try and find peace within yourself before the expectations become too high. Many people look up to those who have a rare talent. The respect of life is to understand that those who possess the mark for creativity – and be given the space they need to continue; so do unto others as you would like others to do unto you. All of us can understand that; no need to smother the gifts of enlightenment. Respect before admiration – the boundaries of psychology.

Unified Field Destiny

Destiny: matching of time and place
"Time: hour, overall view;
first hand, left cerebrum;
second hand, right cerebrum;
seconds: tens, left cerebellum;
ones, right cerebellum."

"Dates: same as time;
year, Number of Occurrences."

Evolution: Biological Destiny.

Sequence Differing

When two particles collide, they are represented by a decimal point.

Numbers

Every number we see is an exact energy occurring within the universe.

Unified Field Destiny Sequence Differ

"Sequence Differ using time equivalents where the third number is added,
subtracted,
multiplied, and divided in a duplication ratio – creativity equal to
the total velocity spin of the model theory 8, 4, 2, the dimensions of the Planck
scale."
$$11{:}26 - 11 \times 2 = 22 - 6 = 16 \div 8 = 2 + 4 = 6$$
(The boundaries of twenty-six)

Bibliography

ESP: The only way to believe is $2c^3$.
Fission: Moving objects that emit gravitational waves.
DNA/RNA: Einstein's prediction.
"Let every idea inspire the next, Indirect Physics as education."
Sociology: Influence is Everything.
"The art of science is the idea; creativity feeds it."
People: Love is ESP.
Mind: Consciousness is equal to love.
"Grand Unification equals 0.00000016 squared twice."
Energy Transversal: Every idea is relevant.
"A scavenger hunt of clues may be exactly what we need."
"Selfishness is only for those who have no faith."

Simplified Scientific Literature: Simple, but subtle.
Rolling Stone: When the community isn't gathering, unpublished data shuttles from one lab to the next, circulation opinions with and beyond the specialties of virologists, immunologists, chemists and cellular biologists. Notebooks are now routinely opened at conferences, a move unthinkable 10 years ago. There is even a proposal afoot to make research papers available online before they are published, threatening a once sacrosanct scientific tradition.
"Once ideas past the test of SSL – the breakthrough would be to write about it."
"Every idea may indirectly feed a resolution."

Freud: Sometimes, the ego blocks the most important information of the subconscious.

Energy Transversal Theory: All ideas come from the physical self.

A penny a day doubled for 31 days equals ≥ 10,737,418.47, total ≥21,474,836.47.

Skeptics: I need to prove a few wrong.

"Amway is not a pyramid, the distribution of products and services has no structure, only Cluster ESP – the tying of emotional dreams, from one person to the next. That is why MLM is so difficult to comprehend."

"Every business, outside of Multi-Level Marketing, is a pyramid. For each business, you work for the owner, who makes all the money. If he likes you, he rewards you with a paycheck, the value. In the Amway Corporation, you are the business owner. You call the shots, within the code or ethics set forth by the corporation. The Cluster ESP, or duplication, can look like anything,

any shape, animal, you name it. It is not limited to anything, other than to people, which is unlimited as far as eyes can see. You can even make more money than your sponsor, because you break-away from him or her when you reach a certain level, where your sponsor's percentage lowers when the level is reached. It has happened, and will continue to. Astronomically, you do not need to sell one product to make your fortune."

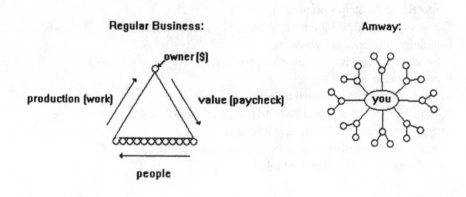

"Amway is the <u>American Way</u>, working smart."

"Art is automatically copyrighted upon completion, all rights reserved."

And the Band Played On. Home Box Office (HBO). Time Warner Entertainment Co., 1993.

Cline, David B. "Low Energy Ways to Observe High-Energy Phenomena." *Scientific American.* September 1994, 40–47.

Crawford, Henry J. and Carsten H. Greiner. "The Search for Strange Quark Matter." *Scientific American.* January 1994, 72–77.

Gardner, Martin. *On the Wild Side: The Big Bang, ESP, the Beast 666, Levitation, Rainmaking, Trance-Channeling, Seances and Ghosts, and More.* Buffalo, New York: Prometheus Books, 1992.

Isabel Briggs Myers w/ Peter Myers. Gifts Differing. Dell Publishing. 1980.

Guterl, Fred. "Keyhole View of a Genius." *Scientific American.* January 1994, 26–27.

Horgan, John. "Particle Metaphysics." *Scientific American.* February 1994, 96–99, 102–106.

Junod, Tom. "Montana Fading Out." <u>GQ</u>. September 1994, 264, 266–269.

Kirshner, Robert P. "The Earth's Elements." *Scientific American.* October 1994, 59–65.

Mathews, Laura. "Girls Best Friend." *Glamour Magazine.* July 1994, 98.

Periodic Table of Elements. *World Book Encyclopedia.* 1982 ed. Childcraft International Inc.: 172–173.

Sullivan, Robert. "In Search of the Cure for AIDS." *Rolling Stone.* April 7, 1994, 56–58, 59–62, 64–65, 85.

Will, Clifford M. *Was Einstein Right?: Putting Relativity to the Test.* New York: Basic Books, 1986.

Weinberg, Steven. "Life in the Universe." *Scientific American.* October 1994, 44–49.

Van Andel, Jay and Richard M. Devos. 1959. <u>Amway Corporation</u>: Multi-Level Marketing and the Process of Duplication.

Unified Field: time elapsed, year and a half.
"Honesty is the best policy."

"A 'no' could mean 'yes' to someone else."
AIDS
"If the AIDS virus was difficult to find through our tests, then, though it seems, those instruments of measure, somewhere – indirectly, has already killed our virus."

Why does it did not survive under certain energies or frequencies – vibrations, 'moving objects emit gravitational waves' may be the resolution – light.

"AIDS is what we aim for."
"The goal is to keep it within our sights."
AIDS Resolution Theory: Collision of particles vs. the virus.
"It attacks cells, so why not attack it."
–My military.
(Why is it so difficult to trace? Is it another penicillin, idea thrown away.)
Indirect Gartish Language
"'ESP and AIDS Equivalents' is equal to 'Terms and Definitions.'"
"Everything is indirectly equal."

Division Point

1. Hormonal forefront
2. Highest point for visualization, hearing and speech
3. Inner child
4. Transfer of creativity
5. Replacement of memory, both physically and mentally
6. Stage before sleep
7. State of mind before sex
8. Emergence of all ideas
9. State of mind during exercise
10. Love
11. Highest level for transfer of ESP
12. ESP Transcription
    1. The body
    2. The mind
    3. visual recognition
13. Expression of sensing, intuition, thinking, and feeling
14. Humor (superego); physical gestures – structures (ego); emergence of dreams (id) into the visual cortex
15. State of bad breath
16. State of boredom
17. State of hunger
18. Life
19. Collapse of consciousness
20. State of mind everyone wants
21. When dreams become reality
22. When drawings become inspiring
23. The zone

Introspect

GROWING UP, I noticed major differences between people. These differences made me curious. I needed to answer for myself why brothers and sisters fought – and why it only happened when they lived together. There seemed to be some force that linked people and aggression – even without words. I even remember watching a few cave movies when I was little. I quickly turned to my father to ask why these noncommunicating hairy creatures that were supposed to be versions of ourselves a long time ago would become angry when they could not speak English. There must be an answer, I said to myself. Thus began my curiosity.

God: something complex as an existence – and his infinite ability becomes difficult to comprehend. The easiest example would be to ask children what gender they thought was God's. The response would be the simple ability to fathom an existence as being gender-specific. So what I have done is to define gender as a math form.

And here we are. Retrospectively, we all are one step away from "breaking down." Anger or violence can be found in all of us. The hint is knowing when and where as a decimal point.

"Within fusion, another universe exists."

BOGART
Structural Psychology

MYERS-BRIGGS TYPE INDICATOR

(poster)

 INFP INFJ INTP INTJ

 ISFP  ISFJ  ISTP  ISTJ

 ENFP  ENFJ  ENTP 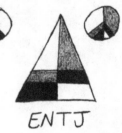 ENTJ

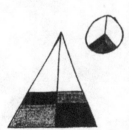 ESFP  ESFJ  ESTP  ESTJ

Indirect Gartish Language

Notes

Pattern Cycle Psychology

"The demonstrated difference between the male and female, or versus another, in a psychological point-of-view – to understand there is a structure to benefit the difference."

- In the world of science, it may be possible for one to combine all elements of life, close enough in origin, to consider itself a phenomenon for others to create their own ideals for further study. In other words, the new beginning of such a "cube" of light, known as extrasensory perception, i.e., indirect physics.

Reminder: The male pattern and the female pattern work in opposite directions.

Superconducting Super Collider

"A huge 11 billion dollar physics project of an underground oval about 54 miles in length to collide subatomic particles in order to calculate their structure. It was of equal brainwash and headaches that was canceled by the House of Representatives in the fiscal year of 1993."

- A challenge physicists say before them; even so, all sciences must once agree.

The Big Bang Equation

"Classical Limitations of the Atom: 85/58."

- Creative math: to see both points of view.

Atomic Equation

"This equation presents physics with the energy
levels for a unified field. The transversals
of geometrical configurations allow us to apply
it to any math of choice: the Planck scale."

Atomic equation: the atom equation:
The velocity of light squared, multiplied
by pie, multiplied by radius, then
divided by the negative space, 16.

Each level and infinite mass can provide us
with a place to start. Finding the errors which
hold us from actual uniform in practicality
can correct the periodic table of elements.

Listing of Pi

"These new pies can give us the difference needed
to substantiate errors of preference for a unified
field for Mathematical World Peace: When we not necessarily
all get along, but instead, when we all agree."

Old: 3.1415927
New: 3.1435836
3.1464266
3.1480152
3.1484205

Projection ESP Pathologies

"Projection ESP Pathologies (PEP) is the study of same sex comparisons. Pathologies for ESP (PE) or are for opposite sex comparisons. These are PEP comparisons."

Black vs. Black – Complete Cancellation as Stress
(2 + 2 = 2: reduces as gray function.)
Black vs. Gray – 2/3 Takeover
Black vs. White – Total Takeover
White vs. White – Incomplete Cancellation
(Only math share.)
White vs. Gray – ½ Takeover
Gray vs. Gray – Complete Function
Gray vs. Black – 1/3 Influence

Black = 2
Gray = 1
White = 0

- These are ideals for study to see influence in both psychology and sociology that can be later induced for creative persistence and intrigue, which can only be found if these shares of individuals are actually shared upon the same energies – such as the same idea or goal.

(The study of psychology will only be denied unless, of course, we do see the firsthand share of pathologies that give us life. These pathologies can be seen any time someone agrees or understands one another. These pathologies are classified as the peace pathologies. What goes on in someone's mind as influence [PEP, 21: Sociology; and PE, 12: Psychology]; so, as to say, the same sex is what defines sociology, and the opposite sex is what defines psychology.)

Pathologies for ESP

"The Half Complete function stated in Pathologies for ESP is known as the Submissive Level Function: Half Complete as an opposite sex Submissive 'Type' Function of study, and what 'types' become submissive during interaction of opposite sexes."

$$ESP = 2c^3.$$

The World Peace Idea

"To combine visual stimulus with structures to
invite a world of imagination to the
Myers-Briggs Type Indicator. And to explain
to the world that world peace is a dream of
all people getting along in an educated world.
But the reality is world peace happens when
we all understand one another."

This is my attempt to capture the understanding, that we all can understand one
another. Points of view separate the differences between us all: these are those
points of view.

Standard Model

The electron has been found to have three quarks.
The proton has also been found to have three quarks.

What has not been found is how many quarks per neutron.

Predicted: two quarks per neutron. Each quark conjoining with the square sides
touching flush with each other. The neutron is both positive and negative quarks
fused together to equal a neutral charge.

God Particle
200,000 mps

Moving Objects Emit Gravitational Waves

"If you were to look at the example of the planet Venus
and what the satellite Magellan picked-up when
it measured the gravitational field due to its topography,
you would see that Venus is a moving object that
emits gravitational force fields."

MOEGW.

MOEGW
Frequencies of gravity of a moving planet.
– General Relativity

ESP Summary

I do not necessarily believe in telepathy as the way that society perceives it to be, but I do believe ESP is a legitimate science with standards that work. I was only being creative to the imagination of ESP as certain standards expressed in my indirect physical series. So to discredit what my views are without knowing how I truly feel may be biased and without actual knowledge of the matter at hand. And also, with my math, it may be possible to prove that ESP has standards but not necessarily the way it has been perceived for many generations.

"Energies do share at levels not construed as being legitimate."

Difference

Men: they see the decision first then they try to control it physically.

Women: they see the physical aspect first then they try to control it emotionally.

Men: dominated by the ego.
Women: dominated by the superego.

<u>Endless Theory</u>

"A Unified Field, Gravity, and Theories of
Everything (TOE) will inevitably use
a sequence of Eepo."

Whenever adding in physics, never carry the ones.

(Eepo: Do not carry the ones.)

Copyright 10/6/01

<u>Garth</u>

"The only mass that can occupy another mass
is negative space itself."

(Bogartium)

-10/16/01

Afterword

THE REALM THAT I see certain ideals that may be complicated are often very simple, thus the name of the book _Simplism_. With rational thought and pure thinking, I hope that the scientific community will give me a chance to express some of my own ideas.

Ever since 1992 and being nineteen years old, I have been working on ideas to promote my venture. And since writing this book when I was twenty-one in 1994, I have come to find some of my ideas correct and some fallible. But all in all, this book was to establish how I perceive my beliefs and convictions.

When my father taught me the Myers-Briggs Type Indicator, established at the University of Florida, he knew it would be a great tool for understanding people. He was absolutely correct. And ever since, I have found a great love for understanding personality types.

In understanding my ideas, you must first understand that I had no idea how to write a book in the first place. I actually stopped writing "Terms and Definitions" and thus began writing "ESP and AIDS Equivalents." I guess my mind was wandering aimlessly.

I also must say that moving objects do emit gravitational waves, as stated by Einstein's prediction. But the rest of the book is a <u>scavenger hunt of clues</u>. Give the ideas a chance and the privilege may be yours: interpretation is the key.

Index

A

absolute time, 14, 19
AIDS, 51-52, 91-92
   crisis, 84
   personality types and, 55
   resolution theory, 92
Amway Corporation, 90
amygdala, 67
angst. *See* depression
arrows, 42, 68
assault, frontal, 78
atomic table, 25
atoms, 53
   boundaries of, 53
   deviants of, 27
   equation, 104
   gravity and, 17, 45
   measurement of, 28
   SSC in resolving secrets of, 15
   See also elements
attraction, 17, 24, 45, 53-55
   chemical, 54
   of mass, 53
   positive, 34

awareness
   emotional, 71, 74
   physical, 72

B

balance, 31, 34, 72
barriers, 23-24, 34, 73
belief, 85
Bell curve, 17
big bang theory, 8, 28, 37-39
biological states of mind (BSM), 47, 58-59, 84
black hole, 33, 38-39, 60
Bogartium, 27, 38, 110
boundaries, 26-27, 31, 35, 40, 44-45
   of the atoms, 53
   negative space and, 26
   safe differential, 57-58
BSM. *See* biological states of mind

C

calcium, 54
calculations, 16, 28, 31

calculator, scientific, 28
cancellation, 19-20, 28, 30, 105
  age, 20
CDP. *See under* developmental patterns:
    conscious
cerebellum, 56, 59, 65, 67, 69, 71-72
cerebrum, 59, 66-67
  left, 57-58, 77, 87
  right, 56, 58, 87
charges
  atomic, 24-25, 53
  neutral, 40, 106
  positive and negative, 17, 24, 29
chemistry, 30, 43, 51-52, 62-63
chloride, 54
collisions, 16, 39, 41, 43
common law, 37
common sense, 15-16, 37
communication, 26, 34, 54-55
  barriers, 27
  language, 26
  written, 34
conception, 42, 47
connection, point of, 41-43
conscious collectivity, 19
conscious mass, 14
consciousness, 54, 56, 58-59, 89, 93
creative math, 28, 39, 47, 60, 103
creativity, 13, 47, 51-52, 86, 89
  transfer of, 71, 93
creativity theory, 52
creator, 35, 70, 85
Creator. *See* God
Curie, Marie, 85
curiosity, 26, 62, 69, 73

D

Darwin, Charles, 85
depression, 84
destiny, 19, 30, 32, 39, 45, 87
  biological, 87

human, 19-20, 32, 47, 53
  unified field, 87
  unity field, 87
developmental patterns, 58, 60
  conscious, 56-57, 59, 84
  subconscious, 56-57, 59
deviants, 27, 34, 70
differential ratio, 41
Discovery Channel, 38
Division Point, 57, 59, 62, 71, 93
dream pattern. *See under* developmental
    patterns: subconscious
dreams, 59, 72, 93
duplication, 14, 25-27, 90
  law of, 25
duplication ratio, 29, 87

E

earth, 32, 56
education, 18, 24, 55, 65
ego, 68-70, 89, 93, 108. *See also* id; superego
egotism parallel, 66, 69
Einstein, Albert, 13, 85
  on absolute time, 14
  energy and, 27, 71
  on imagination, 36
  prediction of, 89, 111
  theory of relativity, 17
electromagnetism, 28, 45, 54
electromagnetivity, 14, 29, 35, 38, 40, 70
electrons, 25, 31, 40, 42, 45, 54, 106
element numbers, 27-28
elements, 15, 26-27, 34, 54
emotions, 55, 59, 69, 73
energy, 15-16, 18, 24, 29, 33-34, 46
  calculation of, 27-28
  fields, 18
  gravity and, 46
  high, 37
  infinite, 34, 37
  infinite mass, 27-28, 33, 35-36, 43

levels, 104
loss, 40-41
of man, 36
numbers and, 87
patterns, 19
transfer of, 77
energy foundation (EF), 34-35, 61
energy pattern theory, 19
energy transversal theory, 47, 69, 89-90
enlightened mass. *See* mass, enlightened
enlightened mass theory, 33
equation
big bang, 5, 21, 37-38, 103
center, 53
creative, 19
energy-loss, 40-41
fusion, 34-35
equilibrium, 44-45
ESP
cluster, 90
reversal, 61, 77
ESPN, 38
evens, 15-16, 19, 31
everything, theory of (TOE), 7, 109
exercise, 71, 93
expectation, 86
extrasensory perception (ESP), 58, 84, 108
cluster, 90
emotional, 73
projection, 86
protective, 73
reverse, 60-62
extroversion, 72
extroverts, 67, 69

F

Faraday, Michael, 43, 70
feeling, 60, 67, 73, 93
feeling individuals, 69, 73, 75
female pattern, 68, 103. *See also* male
pattern

fight or flight, 52-53
frequency, 45-46
frontal lobe, 67, 78
fusion, 34, 62, 97

G

garth, 27-28, 38-39, 110
Gifts Differing, 67, 91
God, 19, 35, 85, 95
gravitational waves, 13, 17, 45-46, 89, 111
gravity, 7, 17, 28-29, 35, 43, 45-46, 70
calculation of, 70
electromagnetivity and, 38, 70
force before, 44
highest range of, 34
pull of, 32
sun and, 34

H

Hanks, Tom, 78
helium, 34
hemispheres of the mind, 14-15, 40, 57, 59,
66, 74
highest mass, 14-15, 34-35, 41, 43
Hubble Space Telescope, 56
hydrogen, 34, 54

I

id, 68, 93. *See also* ego; superego
ideals, 23, 103, 105, 111
imagination, 7-8, 13-14, 36
Indirect Physical Series, 5, 7, 83, 108
indirect physics, 17, 33, 65, 70, 89, 103
infinity, 14
inner child, 71, 93
integers, 58-60, 62
interaction, 38, 44, 55
introversion, 72, 77
introverts, 67, 69, 72

intuition, 7, 35, 73, 93
intuitives, 67, 69, 73, 75, 77

J

Jefferson, Thomas, 76
judgmental individuals, 51-52, 67, 69, 74-75
judgments, 66, 69, 73-74, 77

L

laughter, 71
light, velocity of, 8, 16, 27, 31, 33, 36, 40-41
line, 45-46

M

male pattern, 68, 103
mass, 7, 15, 20, 28, 33, 35-36, 38
  atomic, 27
  captured, 28-29
  ceased, 33
  cubed, 29, 53, 69
  enlightened, 33
  highest, 14-15, 34-35, 41
  opposite, 17
  quark, 18-19, 30, 36
math, 14-15, 17, 23, 28, 47, 54, 62
  creative, 27-28, 103
  derivation of, 16
  purest energy in, 30, 39, 47, 60
  standard, 18
  theoretical, 23
mathematics. *See* math
midbrain, 66-67
miscalculation, 15-16, 20, 29, 41
model theory (8, 4, 2), 25-28, 30, 43, 87
Montana, Joe, 17
Myers, Isabel Briggs, 68, 91
Myers-Briggs Type Indicator, 27, 54, 68,
  83, 111

N

National Institutes of Health (NIH), 51
nature, physical, 52, 72
negative attitude, 24
negative space, 26-27, 34-35, 40-41, 44-45,
  77, 110
neutral charge, 24, 40, 54, 106
neutrals, 24-25, 40
neutron, 25, 31, 39-40, 44, 106
Newton, Isaac, 43, 54
nonmechanical subtlety, 7, 28, 39, 43, 70, 86
number 16, 26, 77. *See also* negative space
numeracies, 19, 30, 35

O

optimism, 24, 55
orbits, 26, 31-33, 38-39

P

paradigm, 28, 41-42, 70
  pattern of, 43
  trilateral triangular, 28
paradoxes, 32
parallel, submission, 66, 69
particles, 28, 31, 38, 43-44
  collision of, 43, 92
  distance between, 45
  mass of, 43
  subatomic, 16, 43, 103
pattern, conscious physical, 14-15
pattern cycle psychology, 103
patterns
  developmental. *See* developmental
    patterns
  dream, 56, 59
perception, 60, 67, 69, 74, 76, 83
periodic table of elements, 14-15, 19, 26,
  91, 104

periodic time, 20
personalities, 19, 55
  destructive, 84
  differing, 56, 59, 66
  types of, 67
personality percentages, 56
personality typing, 55, 67
pessimism, 24
physics, 7, 13, 16, 24, 34, 37, 54
  law, 33
  particle, 44
  psychological, 25, 35, 53, 85
  quantum, 7, 32
pi, 38-39, 104
Picasso, Pablo, 53
Planck scale, 28, 41-42, 44, 87, 104
planets, 38, 40, 46, 60
points of view, 103, 106
Pollyannaism, 54
positive attitude, 24
predictions, 18, 47
Projection ESP Pathologies (PEP), 105
proton, 25, 31, 33, 40, 42, 44, 106
psychology, 25, 29, 65, 72, 86, 105
  indirect physical, 77-78
  pattern cycle, 103
  sociology and, 76-77, 105
  structural, 72
  symbol, 72

Q

quantum mechanics, 32, 45, 83
quantum physics. *See under* physics:
    quantum
quantum time, immeasurable, 39
quark, 7-8, 18-19
  ceased mass of, 33
  equation, 30
  in model theory (8, 4, 2), 25
  number of sides, 44-45

in purest energy, 30
in time travel, 32
velocity spin, 31

R

radius, 28, 104
ratio, 25, 29, 31, 76
  2 to 1, 25, 30, 42-43, 76
  differential, 41
  duplication, 87
reality, 17, 106
  of life, 59
relativity, general theory of, 17, 32, 40, 45,
    83, 107. *See also* Einstein, Albert

S

SAD. *See* social violence and domestic
    abuse
San Francisco 49ers, 17
Scientific American, 38
SDP. *See* subconscious developmental
    pattern
sensing, 54, 73, 93
Sensing individuals, 67, 69, 73, 75
sequence differ, 31, 36, 87
simplicity, 14-17, 23, 27, 58, 83
simplified scientific literature (SSL), 52, 70,
    89
social science, 63
social violence and domestic abuse (SAD),
    76, 83
sociology, 25, 29, 76-77, 105
sodium, 54
spectrums, 7, 56, 71
strong force, 28, 42, 44
subconscious, 47, 55-56, 58-61, 89
  reasoning, 54
subconscious developmental pattern (SDP),
    56, 59

sun, 34, 60, 62
Superconducting Super Collider (SSC), 15-
    16, 103
superconductivity, 16
superego, 68, 93, 108
surface area, 40, 42
symbol theory, 68-69

T

temporal lobe, 67
Texas Instruments Scientific Calculator TI-
    30 III, 28
thalamus, 67
thinkers, 73, 75
thinking, 67, 69, 73, 93, 111
thought pattern, 13, 56
time skip, 33
Time Travel, 20, 32-33, 39, 44-45
time travel, equation, 32

U

unification, 14-16, 31, 42
unified field theory, 14, 19, 28, 43, 70, 104
universal truth, 18-20
universe, 18, 20, 26-27, 43, 85
    collapse of, 39
    creation of, 20, 25
    overall theory about the, 37
    parallel, 19-20

V

velocity, 8, 16, 26-27, 31, 36, 41
visual cortex, 67, 93
volume, 27-28

W

World Book Encyclopedia, 14

Z

zeroes, 45